Andrey Belkin

# World Modeling for Intelligent Autonomous Systems

Karlsruher Schriften zur Anthropomatik
Band 30
Herausgeber: Prof. Dr.-Ing. Jürgen Beyerer

Eine Übersicht aller bisher in dieser Schriftenreihe
erschienenen Bände finden Sie am Ende des Buchs.

# World Modeling for Intelligent Autonomous Systems

by
Andrey Belkin

KIT Scientific Publishing

Dissertation, Karlsruher Institut für Technologie
KIT-Fakultät für Informatik

Tag der mündlichen Prüfung: 16. Dezember 2015
Referenten: Prof. Dr. Jürgen Beyerer, Prof. Dr. Tamim Asfour

**Impressum**

KIT Scientific Publishing

Karlsruher Institut für Technologie (KIT)
KIT Scientific Publishing
Straße am Forum 2
D-76131 Karlsruhe

KIT Scientific Publishing is a registered trademark
of Karlsruhe Institute of Technology.
Reprint using the book cover is not allowed.

www.ksp.kit.edu



Print on Demand 2017 – Gedruckt auf FSC-zertifiziertem Papier

ISSN 1863-6489
ISBN 978-3-7315-0641-6
DOI 10.5445/KSP/1000066938

# Contents

# List of Acronyms and Symbols

## Acronyms

| | |
|---|---|
| AMROS | Autonomous Multisensoric Robot for Security Applications |
| ARMAR | AnthRopomorphic Multi-Arm-Robot |
| DARPA | Defense Advanced Research Projects Agency |
| DFG | Deutsche Forschungsgemeinschaft (German Science Foundation) |
| DM | dynamic model |
| DoB | degree-of-belief |
| FPGA | Field Programmable Gate Array |
| GISD | generalized integrated squared difference |
| GM | Gaussian mixture |
| IES | Lehrstuhl für Interaktive Echtzeitsysteme (Chair of Interactive Real-Time Systems) |
| IOSB | (Fraunhofer) Institute of Optronics, System Technologies and Image Exploitation |
| IPDA | Integrated Probabilistic Data Association |
| ISD | Integrated Squared Difference |
| JDL | (U.S.) Joint Directors of Laboratories |

| | |
|---|---|
| JIPDA | Joint Integrated Probabilistic Data Association |
| KF | Kalman filter |
| KL | Kullback-Leibler |
| LIDAR | light detection and ranging |
| LDQ | least discernible quantum |
| MAP | Maximum A Posteriori |
| MDL | minimum description length |
| MHT | Multiple Hypothesis Tracking |
| NEST | Network Enabled Surveillance and Tracking |
| NN | Nearest Neighbor (algorithm) |
| OPASCA | opto-acoustic scene analysis |
| OWL | Ontology Web Language |
| PDA | probabilistic data association |
| pdf | probability distribution function |
| p.d.f. | probability density function |
| PF | particle filter |
| PK | prior knowledge |
| PM | Progressive Mapping |
| $PM^2$ | Progressive Mapping with Prior knowledge Matching |
| p.m.f. | probability mass function |
| Robocup | Robot Soccer World Cup |
| SFB | Sonderforschungsbereich (Collaborative Research Center) |
| SLAM | simultaneous localization and mapping |
| WoI | world of interest |
| WC | weak-continuous |
| WD | weak-discrete |
| WSD | weak squared difference |

## General Notation

| | |
|---|---|
| $a, \sigma, \ldots$ | scalars (small, italic) |
| $\vec{\boldsymbol{a}}, \vec{\boldsymbol{\sigma}}, \ldots$ | vectors (small, italic, bold, vector head) |
| $\boldsymbol{A}, \boldsymbol{\Sigma}, \ldots$ | matrices (capitalized, italic, bold) |
| $\{A\}, \{\Sigma\}, \ldots$ | sets (capitalized, italic, braces) |
| $\|\{A\}\|, \|\{\Sigma\}\|, \ldots$ | number of set elements (capitalized, italic, braces, vertical bars) |
| $A, \Sigma, \ldots$ | random variables, denoting attributes (capitalized, italic) |
| $\bullet_k$ | notion "•" at time step $k$ (index $k$, small, italic, e.g. $a_k$) |

## Symbols

| | |
|---|---|
| $p(A) = p(a)$ | normalized probability distribution, i.e. probability distribution function (pdf) of $A$ over values $a$ |
| $p(A) = p(a_i)$ | probability mass function (p.m.f.) of $A$ over $A$'s possible states $a_i$ |
| $p(A_1, A_2, \ldots, A_N)$ | joint probability distribution function of $N$ random variables |
| $p(A\|B)$ | conditional probability (probability of $A$ given $B$) |
| $\text{var}(A)$ | variance of $A$ |
| $\text{cov}(A_1, A_2)$ | covariance of $A_1$ and $A_2$ |
| $\boldsymbol{\Sigma}_{\boldsymbol{A_1},\boldsymbol{A_2}}$ | covariance matrix of $A_1$ and $A_2$ |
| $\sigma$ | standard deviation |
| $\mu \equiv E[A]$ | expectation of $A$ (differs from observation $\vec{\boldsymbol{\mu}}$ in KF or $\vec{\boldsymbol{\mu}}$ label in weak distance calculations) |
| $p(A) = N(\mu, \sigma; a) \equiv N(\mu, \sigma)$ | univariate Gaussian (normal) distribution |

| | |
|---|---|
| $p(A) = N(\vec{\boldsymbol{\mu}},\boldsymbol{\Sigma};\vec{\boldsymbol{a}}) \equiv N(\vec{\boldsymbol{\mu}},\boldsymbol{\Sigma})$ | multivariate Gaussian (normal) distribution |
| $GM\left(\{w_k,\vec{\boldsymbol{\mu}}_k,\boldsymbol{\Sigma}_k\}_{k=1}^{K};\vec{\boldsymbol{a}}\right)$ | multivariate Gaussian mixture (weighted sum of $K$ Gaussian distributions with $w_k$ weights) |
| $t_k$ | time step $k$ |
| $D_\bullet$ | aging threshold for notion "•" (i.e. $c$ - creation, $r$ - reconfirmation, $d$ - deletion) |
| $F_\bullet$ | aging factor for notion "•" |
| $T$ | temperature or class of probe functions |
| $\Delta t$ | time quantum |
| $\vec{\boldsymbol{s}}_k$ | state at time step $k$ |
| $\tilde{\vec{\boldsymbol{s}}}$ | (unobservable) true state of the system (refer to KF) |
| $\boldsymbol{F}_k$ | propagation matrix (dynamic model for a projection of the state into future, refer to KF) |
| $\boldsymbol{B}_k$ | control-input model matrix (refer to KF) |
| $\vec{\boldsymbol{u}}_k$ | control vector (refer to KF) |
| $\vec{\boldsymbol{w}}_k$ | random process noise (refer to KF) |
| $\boldsymbol{Q}_k$ | covariance of the random process noise (refer to KF) |
| $\vec{\boldsymbol{\mu}}$ | observation (refer to KF, differs from expectation $\mu$) |
| $\vec{\boldsymbol{v}}_k$ | observation noise (refer to KF) |
| $\boldsymbol{R}_k$ | covariance of the observation noise (refer to KF) |
| $\hat{\vec{\boldsymbol{s}}}_{k\|k-1}$ | state estimate (refer to KF) |
| $\hat{\boldsymbol{P}}_{k\|k-1}$ | covariance matrix of the state estimate (refer to KF) |
| $\boldsymbol{I}$ | identity matrix |
| $\boldsymbol{K}_k$ | Kalman gain (refer to KF) |
| $\boldsymbol{S}_k$ | innovation (refer to KF) |
| $\langle f,\psi\rangle$ | generalized moments of functional $f$ and probe function $\psi$ |

| | |
|---|---|
| $\delta(x)$ | Dirac delta function |
| $\widetilde{x}_k^i$ | particles, where $i \in 1,\dots,N$ (refer to PF) |
| $\theta_k$ | random process noise (refer to PF) |
| $f\left(\widetilde{x}_k^i,\theta_k\right)$ | evolution function (refer to PF) |
| $\widehat{x}_{k+1}^i$ | particles of the resampling stage, where $i \in 1,\dots,N$ (refer to PF) |
| $y_{k+1}$ | observation (refer to PF) |
| $D_A$ | definition set of the attribute $A$ |
| $H(A)$ or $h(A)$ | information entropy of the attribute $A$ |
| $H_\Delta(A)$ or $h_\Delta(A)$ | pre-discretized information entropy of the attribute $A$ |
| $\Delta_A$ | slice by histogramming procedure of the attribute $A$ |
| $v$ | slice index by histogramming procedure |
| $B(A)$ | Bestimmtheit of the attribute $A$ |
| $r$ | representative |
| $c$ | concept |
| $\{C\}_r$ | selected (during prior knowledge matching) set of representatives |
| $d(r,c)$ | metric that quantifies a difference between $r$ and $c$ |
| $\lambda_s$ and $\lambda_v$ | weighting factors for structural and value distances |
| $\{A\}_r$ and $\{A\}_c$ | sets of attribute descriptions of $r$ and $c$ respectively |
| $d_{\text{s:}\bullet}$ | structural difference metric with a name "•" (e.g. $d_{\text{s:Jaccard}}$) |
| $d_{\text{DoB:}\bullet}$ | distance between two pdfs with a name "•" (e.g. $d_{\text{DoB:KL}}$) |
| $L^2$ or $L2$ | Wasserstein metric |
| $\psi(\vec{\boldsymbol{x}})$ | probe function |
| $P$ | finite-dimensional set of parameters (for a probe function parametrization) |
| $\langle f,\psi\rangle$ | generalized moment of a functional $f(\vec{\boldsymbol{x}})$ and a probe function $\psi(\vec{\boldsymbol{x}})$ |

| | |
|---|---|
| $\delta(x)$ | Dirac delta-function |
| $d_T[f,g]$ | weak distance |
| $w[\psi]$ | real-valued functional denoting the density of probe functions per unit volume |
| $\vec{\boldsymbol{\mu}}$ | characteristic position and a label for a probe function |
| $\Omega_\psi(\vec{\boldsymbol{\mu}})$ | utility function |
| $s_\Omega$, $s_p$ | characteristic scales |
| $\Sigma_{ISD}$ | symmetrized normalized ISD |
| $\Sigma_{WSD}$ | symmetrized normalized WSD |

# 1

# Overview

## 1.1 Autonomous Systems

### 1.1.1 Introduction

From the very beginning of conscious existence, mankind has tried to create tools and routines for its survival. More advanced tools have allowed to protect human beings from environment hazards better and to get resources more efficiently. This has led to an increase of the free time amount, and this free time was in turn spent for improvements of tools and routines. The loop of *"perfection of tools and routines ↻ more free time"* was established.

Improvements in work facilitation led to dreams of systems that would take the whole work from humans over, e.g. magic items or golems, which carry out duties without human supervision. With time passing, the invented tools became semi-autonomous and even partially conscious with limited reasoning. Moreover, recent developments in computer science, especially in robotics, allowed creation of systems that are autonomous over relatively long time periods. Hereby, we denote *autonomous systems* as systems able to perform designed tasks in dynamic environments without continuous human guidance. As we state in [Bel12c], the purpose of such systems is to function stand-alone according to pre-defined strategy, e.g.

protecting areas or rescuing people in hazards. A typical *operation cycle* of an autonomous system in a dynamic environment is presented in Fig. 1.1.

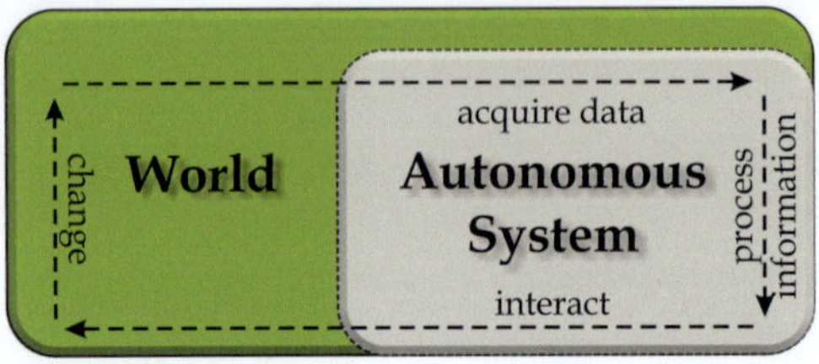


**Figure 1.1:** Operation cycle of an autonomous system.

It is obvious that different autonomous systems are autonomous in different ways with different degrees of autonomy. We define a *fully autonomous system* as a system with abilities to

- acquire information about the surrounding environment,
- function during a planned period of time without human assistance,
- interact with the environment and other systems,
- follow pre-defined and emerging human instructions,
- avoid effects that are harmful for people, property or itself unless it was specified in the design of the system.

The last two statements are equivalent to the *Three Laws of Robotics* of Isaac Asimov [Asi48].

The autonomous task fulfillment and interactions with the environment over prolonged periods of time is challenging. If the environment as well as tasks are complex and diversified, correspondingly complex anticipation and planning are required. Thus, simple immediate responses to the environment changes, though still vital, are no more sufficient.

### 1.1.2 Reactivity and Proactivity

We distinguish two types of interaction with the environment. An immediate reaction to the acquired information, e.g. shifting of center-of-mass during balancing, is called *reactivity*. Reactive behavior allows autonomous systems to immediately respond to environment changes, especially to time critical issues and threats, in a direct manner, as human reflexes do (an extended discussion about this topic is given e.g. in [Ark98] and a multi-robot formation example is presented in [Bal98]). However, the use of reaction-based architectures for solving complicated tasks can frequently be problematic [Lim11]. Complex tasks, such as rescuing people in hazards, require sophisticated anticipation of future changes and situations, thorough decision making and planning. This kind of complex self-initiated rational reaction is called *proactivity*.

The basis for proactive reasoning is a constant *situation awareness* that is defined by Endsley [End95] as

> the perception of elements in the environment within a volume of time and space, the comprehension of their meaning, and the projection of their status in the near future.

This implies a significant extension of the information processing in the operation cycle compared to the reactive behavior. A proactive autonomous system has to thoroughly analyze the available information, understand present situation and predict changes in the environment and itself. Thus, such systems have to be intelligent.

### 1.1.3 Intelligent Autonomous Systems

An unsupervised functioning in an open or hazardous environment for a prolonged period of time demands corresponding cognition, decision making and planning, i.e. intelligence. An *intelligent autonomous system* was defined within the *IBM Autonomic Computing Initiative* in [IBM01], [Kep03] with the following abilities:

- *environment-awareness* – ability to perceive and know the surrounding environment and context;
- *self-awareness* – ability to know system's own components, current status, ultimate capacity and connections to other systems;

- *self-configuration* – ability to configure and reconfigure itself under internal and external changes;
- *self-protection* – ability to protect itself;
- *self-healing* – ability to discover and correct faults;
- *self-optimization* – ability to optimize own working.

It is clear that an intelligent autonomous system has to perform complex cognition and decision making tasks. For example, it has to constantly optimize the sensors' configuration in order to acquire as much relevant information as possible [Ahl02], [Hub09]. Also, it has to identify the information deficit relative to given tasks [Bel12b], notice interesting environment elements and changes [Sch11] and perform task-oriented data-driven exploration [Küh10]. The acquired information has to be stored along with pre-defined information in an efficient and robust way for situation and planning analysis and has to be available to all other subsystems [Bel10], [Bel12d]. A possible solution to these issues can be a memory structure containing a dynamic description of the observed environment (dynamic model) and persistent prior knowledge database, similarly to the short-term and long-term human memories. A self-optimization can be performed with a configuration space analysis [Kai11], learning through cognition inferences and references from humans and other autonomous systems [Jäk11], and extension of the prior knowledge [Kuw13]. The latter can be performed, for example, if the introduction of a new prior knowledge concept would reasonably reduce the size of the dynamic description of the relevant environment [Bel12d]. Dynamic model and prior knowledge represent a *world model* – the complete known information reference about the world with reasoning means about this information. Thus, the world modeling subsystem is the central component for all other subsystems and a basis for all cognition and decision making processes.

# 1.2 World Modeling

## 1.2.1 Introduction

The functioning of intelligent autonomous systems requires constant situation awareness and cognition analysis. The basis for such awareness and analysis is the central memory structure, called world modeling subsystem. It contains a description of the surrounding environment, conformed to prior knowledge, and serves as an information hub for all other subsystems [Bel10]. The information input for the world description can be provided by sensors and cognition modules. In order to clearly understand the place of the world modeling within the autonomous system, it is useful to have a global outline of the subject area, namely, to define relevant realms, domains and their relations.

## 1.2.2 Realms and System Domains

We define autonomous system global structure by specifying distinct *realms* and *domains* (Fig. 1.2).

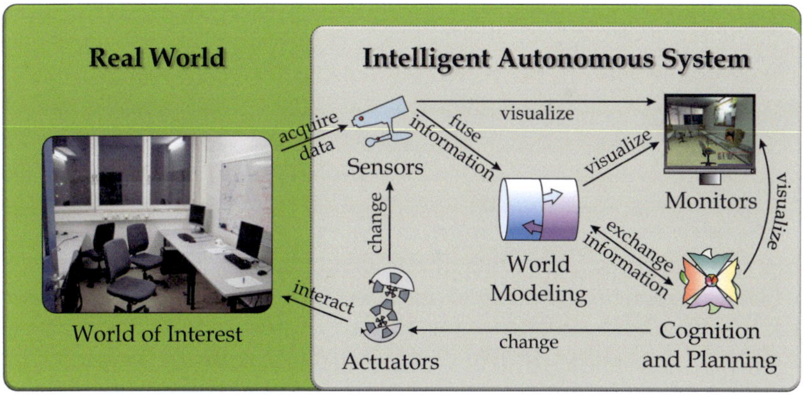


**Figure 1.2:** Autonomous system realms and domains.

We distinguish two realms: the real world and the intelligent autonomous system itself. The real world contains a *world of interest* (WoI) domain, which represents a spatio-temporal section of the surrounding environment relevant to the autonomous system. The autonomous system acquires data about the world of interest by *sensors* and interacts with it by *actuators*. The sensors domain is responsible for the raw data acquisition, pre-processing of data (e.g. 3D spacing of video stream) and converting these data to useful information (e.g. entity's position). The acquired information is directed to the world modeling subsystem, which contains WoI description and serves as a global information hub for all other subsystems, e.g. *cognition and planning* modules. The latter assess situations, plan future actions and give orders to actuators. Some autonomous system components can visualize its own information on *monitors*. Thus, we have 6 domains under consideration, all of them except actuators are related to the world modeling.

The presented domains scheme describes most architectures of intelligent autonomous systems. Before defining the world modeling subsystem in detail, it is vital to introduce basic notions and terminology for relevant domains.

### 1.2.3 Notions and Terminology

The dedicated notions and terminology are introduced in our work [Bel12d] as follows. We refer to the elements of the world of interest as *entities*. An entity can be movable (e.g. person), stationary (e.g. room) or non-material (e.g. alarm). An example of an object-oriented decomposition of WoI relevant to surveillance systems is presented in Fig. 4.8. We call virtual elements of the world model, corresponding to real entities, as *representatives*. Each representative describes a hypothesis about an entity, in particular, there can be several representatives to one entity as a result of ambiguity or clutter. The representatives are populating a dynamic world model, which is a description of the world of interest. While the dynamic world model has to be kept slim (due to processing complexity), prior knowledge back-end has to contain most complete information about the world for giving semantic meaning to representatives and extending them with additional information.

Each attribute of an entity is given by an *attribute description* in the world model. In order to cope with uncertainties, we suggest that such descriptions be probability distributions over possible values. A representative contains a set of attribute descriptions obtained from observations or inferences. The attribute descriptions can be persistent (mostly time-invariant, e.g. person name or table length) and variable (changing over time, e.g. position of a car).

Entities within the world of interest are often related to each other. The corresponding links, called *relation descriptions*, connect representatives to each other. Relation descriptions can be quantitative (e.g. distance between representatives) or qualitative (e.g. fact: one representative is a part of another one).

A consolidation of attribute and relation descriptions of a representative is called *state estimate*, which is a description of the corresponding entity's state conformed to prior knowledge and operational tasks.

Groups of representatives can form complex bounded constellations. We denote a momentary snapshot of the dynamic world model as a *scene* [Ghe08]. A time sequence of scenes, called an *episode*, represents an evolution of the dynamic world model with a given time quantum (Fig. 1.3). Episodes can be valuable, enabling temporal analysis that is vital for many complex inferences.

A qualitative assessment of scenes and episodes allows for semantic interpretations of the constellations of representatives. A result of such assessment can be a symbolic conclusion, called a *situation* – a compact and essential description of the scene at hand. Although many attribute and relation descriptions could be lost here, the simplicity of such description allows for an instantaneous scene interpretation by a human or planning subsystem and results in situation awareness. One of the simplest situations is a qualitative assessment of a representative's state, for example PERSON IS SMILING. More complex situations involve an analysis of relation descriptions (e.g. TWO PERSONS HAVE A CONVERSATION) or temporal inferences (conclusions about prolonged activities, e.g. PERSON IS DANCING). Situations can be inferred from other situations, e.g. if several persons are dancing and several other are drinking beer, the inferred situation can be PARTY. A found situation is assigned to concrete scenes and has to be verified at later scenes.

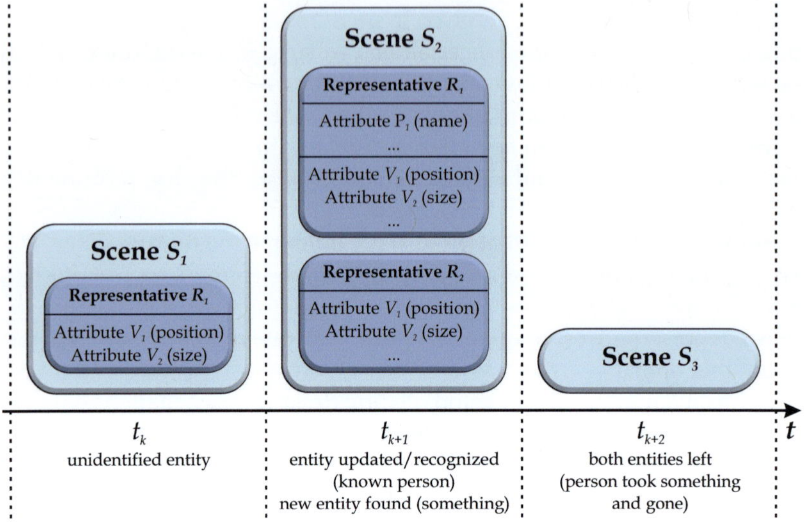


**Figure 1.3:** Evolution of a scene.

This qualitative assessment is discussed in our work [Bel12d]. A connection of relations and situations is presented in Figure 1.4.

We have introduced now the sufficient vocabulary and enough notions about the subject to describe our ideas. However, it would be advantageous to review existing publications within the field of world modeling. The existing works often serve as a basis for new advances, i.e. by defining which methods and concepts exist and which problems and challenges are awaiting our attention. Also, a rational assessment of new ideas can be performed in the context of existing works. Thus, our research of the world modeling starts with an overview of existing solutions and technologies, i.e. with an overview of the state of the art within the field of interest.

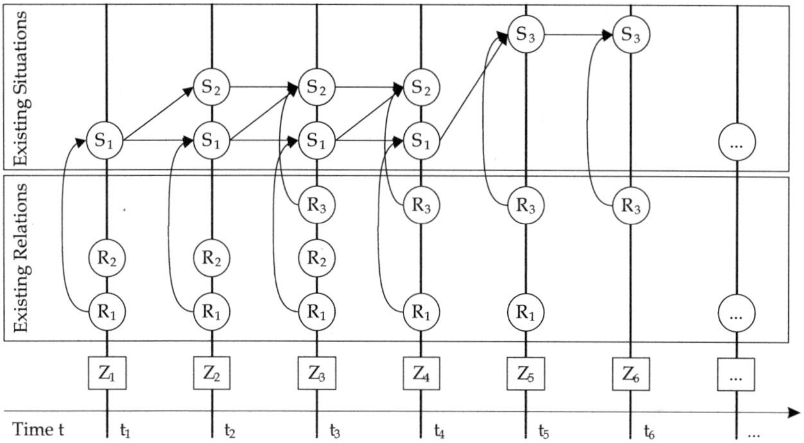


**Figure 1.4:** Connection of relations $R_i$ and situations $S_j$ for each given at time step $t_k$ scene $Z_n$ [Bel12d].

# 1.3 State of the Art

## 1.3.1 Introduction

Due to computational and engineering complexity, modern intelligent autonomous systems are limited to a concrete set of tasks [Bel10]. For example, autonomous vehicles perform exploration, localization and path planning, limiting the world modeling subsystem to geometrical and topological aspects of the environment (e.g. to so-called simultaneous localization and mapping – SLAM). Another example – an autonomous receptionist and a medical assistant robot classify surrounding entities and situations for reactive decision making. In general, we distinguish the following types of autonomous systems:

- stationary platforms – immovable systems for manufacturing, medical assistance, etc.;

- mobile platforms – maneuverable units for military and rescue operations, many bionic systems, etc.;
- systems of simple multi-agents – so-called swarms, multiple miniaturized sensor-actor agents;
- systems of intelligent multi-agents – surveillance systems, football robotic teams, etc.;
- humanoid robots – robots with human-like appearance and behavior for service, entertainment and other purposes.

### 1.3.2 Existing Intelligent Autonomous Systems

#### Stationary Platforms

Stationary autonomous platforms, such as radars, manufacturing robots for production lines (e.g. KUKA robot [KUK08] in Fig. 1.5) or surgery robots as a tabletop system at Duke University [Rog09]), are designed to perform usually one given task.

#### Mobile Platforms

There are many mobile platforms developed for autonomous exploration, tracking, cargo carriage and military operations (e.g. Fig. 1.6). A typical example is one of the first driverless cars *VaMP*, which was able to drive in heavy traffic for long distances autonomously [Dic94]. The *Grand Challenge* [Bue07] in 2005 and the *Urban Challenge* [Bue10] in 2007 organized by the Defense Advanced Research Projects Agency (DARPA) demonstrated that the technology has reached the level of fully autonomous drive of hundreds kilometers off-road by cars, passing narrow tunnels and more than 100 sharp turns, or tens kilometers of urban area obeying traffic regulations while dealing with traffic and obstacles. Another example is the *Legged Squad Support System* [DAR08] designed by Boston Dynamics and DARPA – a legged robot, which represents a packhorse for military operations, with up to 180 kg payload and following a human lead on most of terrains and complex environments autonomously. Yet another example is the Autonomous Multisensoric Robots for Security Applications (AMROS) system

**Figure 1.5:** Stationary platform: KUKA robot hand for autonomous deflectrometric inspection of varnish coating at IES laboratory [Leh11].

(Fig. 1.6b), currently developed at Fraunhofer IOSB [Fra08] – an autonomous mobile robotic system for multi-sensor in-/outdoor surveillance of building complexes [Emt07]. The mobile platforms are not limited to ground vehicles – there is a variety of aerial [Non10], [Seg11] (e.g. Fig. 1.6a) to underwater [Gri02] systems available.

In addition to the "usual" mobile platforms with strong data acquisition and maneuvering abilities, there are platforms designed for advanced cognitive information processing. For example, service robotics is focused on development of museum guides, receptionists and companions (e.g. BIRON mobile platform [Haa04]).

**(a)** Aerial reconnaissance drone;
©Fraunhofer IOSB, photo Manfred Zentsch

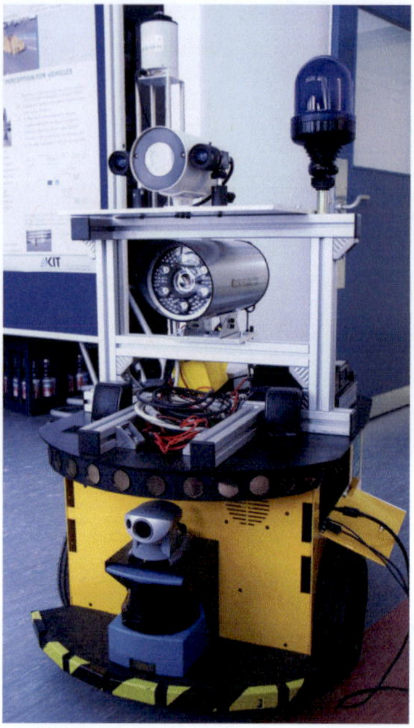

**(b)** AMROS robot for autonomous patrolling and SLAM;

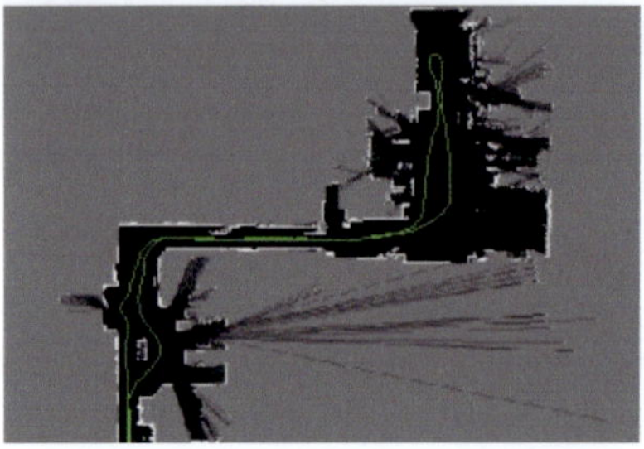

**(c)** AMROS SLAM: reconstructed floor map with robot's path (green) and free space (black).
©Fraunhofer IOSB, courtesy Thomas Emter

**Figure 1.6:** Mobile platform examples.

## Systems of Simple Multi-Agents

We call a multiple-agent system, consisting of many most simple and inexpensive mobile units, as a *swarm*. While each unit possesses only limited abilities, connections between swarm members allow for decentralized hive intellect and behavior. In principle, swarm robotics is suited for cheap design tasks, such as mining, pipes inspection or vermin extermination. Additionally, such miniaturized robots can be applied in micromachinery or be injected into living beings for distributed sensing and delicate manipulations. Notable examples of the swarm robotics are the Swarmanoid [Dor11] and the Kilobot [Rub12] projects.

### Systems of Intelligent Multi-Agents

The intelligent multi-agent systems include e.g. smart distributed surveillance systems. One such system is *Network Enabled Surveillance and Tracking* (NEST) [Bau09b] (Fig. 1.7), developed at Fraunhofer IOSB. It usually contains multiple mobile and stationary components that are complex and distributed. Mobile sensor devices, such as airborne miniature drones (e.g. Fig. 1.6a) or wheeled patrolling platforms (e.g. Fig. 1.6b), are intended to (autonomously) scout the assigned area while stationary sensors (for example, wall-mounted cameras and microphones) constantly monitor the same area. Centralized processing nodes fuse the incoming multi-modal sensory data into a single world model, estimate target attributes and environment situation and project them into the future. Another example is a robot team for absolutely autonomous work and cooperation, e.g. Robocup [Rob08] teams.

### Humanoid Robots

A humanoid robot (android) is one of the most complex and advanced intelligent autonomous system concepts. The ultimate goal in this research area is the creation of an anthropomorphic robot that is similar in appearance, cognitive abilities and behavior to a human. This means, it has to contain all mechatronic components within a human-like body and to use two cameras, microphones and tactile and inertial sensors only. Moreover, most human-like reactive and proactive responses have to be reproduced, as well as presence of the constant situation awareness, robust decision making and flexible learning. At last, it has to perform social interactions with people and other robots and ensure their safety.

The humanoid robots were always the very special goal, like creation of an "artificial self". Notwithstanding, it is still an extremely difficult task, though, modern technologies render this challenge as an "asymptotically achievable" goal. Nowadays, many research institutes and companies are working on the creation of the most advanced androids.

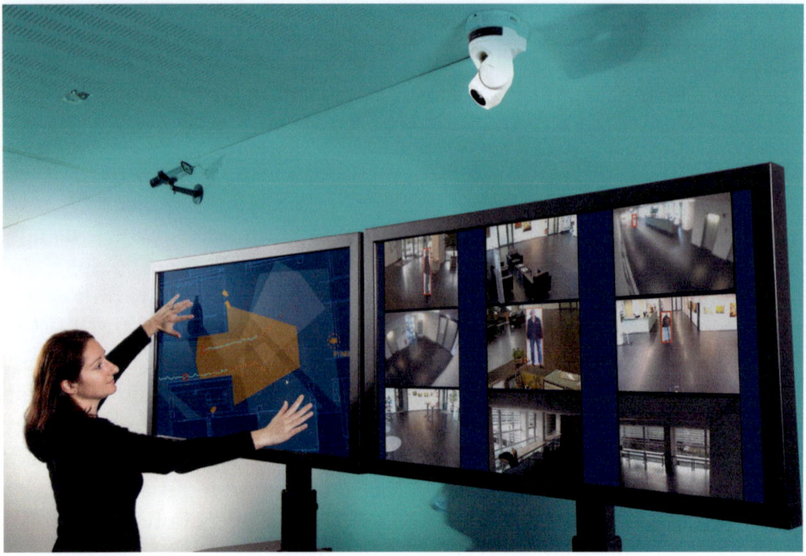

**Figure 1.7:** Cameras and monitors of the Network Enabled Surveillance and Tracking (NEST) system.
©Fraunhofer IOSB, photo Manfred Zentsch

The purposes of the creation and intended application fields of the anthropomorphic robots, however, are various. This naturally comes from a need of a significant outcome from massive investments and complex research and impossibility of creation of a universal intelligent robot equal to human in the nearest future. For example, the United States pay great attention to the possibility in principle of creating of humanoid robots as well as of (para)military humanoid robots. Modern US military programs once focused on autonomous mobile platforms (e.g. already mentioned DARPA Grand and Urban Challenges) have turned their attention to humanoid robots, which shall replace human soldiers and excel enemy forces in future battlefields. The examples for it are DARPA *Autonomous Robot Manipulation (ARM) Robot* [DAR10], Vecna Technologies *Battlefield Extraction-Assist Robot (BEAR)* [Vec05], and Virginia Tech College of Engineering, Robotics and Mechanisms Laboratory funded by DARPA *Cognitive Humanoid Autonomous Robot with Learning Intelligence (CHARLI)* [Lah08] or its Navy version developed with US Naval Research Laboratory – *Shipboard Autonomous Firefighting Robot* (SAFFiR). Another Navy humanoid robot example is *Lucas* and its female counterpart, *Octavia.* Moreover, in order to stimulate the humanoid robots development, DARPA had organized the *Robotics Challenge* 2012-2013. To this end, Boston Dynamics had delivered a hardware platform for all contestants – a bipedal robot *PETMAN.* It is able to move dynamically like a real person, sweat, and execute complex tasks in dangerous, degraded, human-engineered environments. Much of its technology was derived from DARPA's sponsored *BigDog* platform. There is also a space program robot developed by NASA Johnson Space Center – the *Robonaut* [Amb00].

In contrast to the USA, research groups in Asia are focused on humanoid robots with perfect maneuvering abilities – bipedal platforms that can run and dance, play musical instruments and work at bars, express emotions during conversations and help at information centers. The most famous example is the Honda *Advanced Step in Innovative Mobility* (ASIMO) (preceded by robots of *E* and *P* series) [Hon07]. Initially planned to be a helper to people, it was turned, due to enormous complexity of arising challenges, to a platform for sensor-actor research and development. For example, ASIMO can walk and run, recognize moving objects, faces, postures, gestures, its surrounding environment, different sounds and speech. All these abilities are unfortunately limited to simple cases. Another examples are *TOSY Ping Pong Playing Robot* (TOPIO) [TOS10] or *TOPIO Dio* by TOSY Robotics, *Humanoid Robotics Project* (HRP) [Kan08] by Japanese collaboration of government and industry institutions, and Toyota *Violin-playing robot* [Kus08]. There are androids with human-like appearance, mimic and gestures, for example, *Geminoid HI-1* [Nis07] and *Geminoid F, Repliee Q2* [Mat05], and *EveR-2* [Ahn11].

The humanoid robots mainstream in Europe is robots with strong cognitive abilities (e.g. household anthropomorphic robots) or technical platforms for entertainment and research purposes. One of the examples is the *AnthRopomorphic Multi-Arm-Robot* (ARMAR) [Asf06] (Fig. 1.8) created within the Deutsche Forschungsgemeinschaft (DFG) Collaborative Research Center 588 "Humanoid Robots – Learning and Cooperating Multimodal

Robots" [Deu08]. This robot is able to put dishes into a dishwasher or bring apple juice from a fridge upon a speech command. Another example, is the famous *Nao* robot developed by Aldebaran Robotics. Created initially for entertainment, this robot had progressed rapidly and became a technological platform for many research groups over the world, as well as the main platform for the *Robot Soccer World Cup* (Robocup) Standard Platform League (SPL) since 2008. In 2010, Nao performed a synchronized dance at the Shanghai Expo in China. One more example is a universal anthropomorphic platform *REEM* [Tel08] by PAL Robotics.

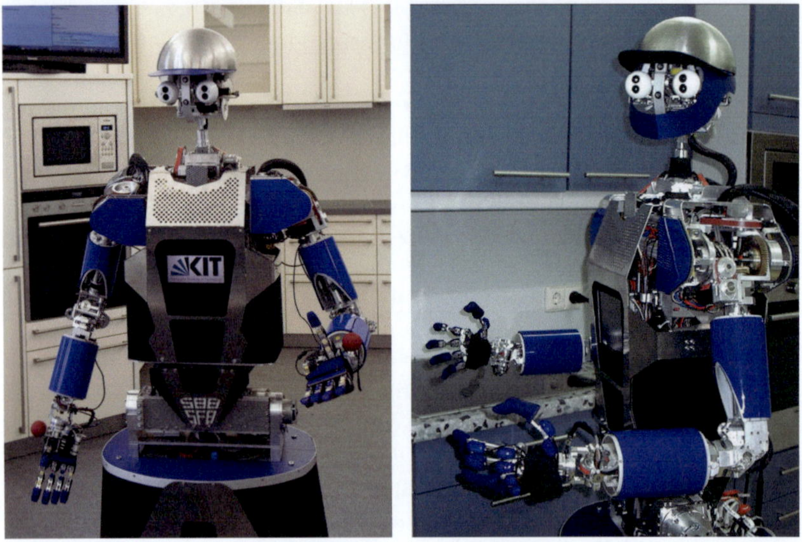


**Figure 1.8:** ARMAR of the 3rd generation.

### 1.3.3 Existing World Modeling Architectures

#### Introduction

Modeling of the world of interest is in essence modeling of heterogeneous information. Formal methods for doing so are based on semantic networks, first-order logic and formal languages [Rus10]. Practical implementations involve ontologies, object-oriented and probabilistic approaches, e.g. discussed in [Iso01], [Sir07], [Cos08], [Ghe08], [Pap08], [Bau09a].

As we mention in [Bel10] and [Bau10a], the proposed approaches in modeling of the environment are domain-specific and usually not extendable to other applications. Even the modeling for mobile platforms is not yet solved, for example, [Fur10] states:

> Although numerous solutions have been developed for modeling the environment of autonomous robots and/or off-road autonomous vehicles, no solution or proposal has yet been published, which is capable of modeling all relevant aspects of urban roads, which are required for autonomous city vehicles safely operating in non-simplified urban traffic conditions ... For example, the developers of the DARPA Urban Challenge winning vehicle BOSS observed that their traffic representation was not sufficient to make intelligent driving decisions compared to human drivers. Furthermore, Junior's developers (2nd place) noticed that their vehicle, and probably any other vehicle competing in this race would not be able to cope with a realistic city traffic environment.

This comes from the fact that modern autonomous systems are challenged by limited possibilities within complex engineering, i.e. by tasks that overwhelm the available computer processing power within time critical scenarios [Bel12b]. For anthropomorphic service robotics, [Bee10] makes the following statement about knowledge management frameworks:

> We investigate autonomous robot control systems that enable robots to perform complex everyday manipulation activities in human living environments. Such control systems are, for instance, needed for autonomous household robots. The design, implementation, and deployment of robot control systems for such complex applications is a challenging and intense programming task that requires powerful software tools. To respond to these needs, a number of middle-ware software libraries that support the development of distributed modular control systems have been developed. These middleware

> systems include ROS [Qui09], Player [Ger03], Yarp [Fit08], and Orocos [Smi08]. There is, however, a lack of powerful software tools that enable programmers to effectively and efficiently implement higher-level capabilities such as learning, knowledge processing, and action planning into the robot control programs to produce more flexible, reliable and efficient behavior.

At the beginning of robotics, there existed modeling systems for very specific applications only (e.g. refer to an overview in [Ang92]). Most of those and even of later works have considered only spatial modeling, e.g. occupancy grid map [Thr03], [Har04], [Emt10]. However, the difference in purposes and worlds of interest has led to a diversity in world modeling. Exempli gratia, [Iso01] has proposed an object-oriented world modeling for simulation of virtual environments and system engineering within the financial domain only. The work [Pap08] has described a dynamic, object-oriented modeling relevant for cooperative vehicles. An analysis of urban traffic world modeling for autonomous cars is presented, for example, in [Ben06] but also limited to static and dynamic obstacles.

Since tasks assigned to robots become increasingly challenging, the modeling subsystems evolve in the direction of growing complexity and applicability. For example, [Die12] discusses a separation of sensory data and world models. [Cau00] considers psychological principles for perceiving, modeling and abstracting information. Nevertheless, there is still no universal knowledge representation and processing platform suitable to cope with modeling of realistic environments. Due to arising application-scope limitations, we will consider existing approaches categorized by intelligent autonomous system types.

### World Modeling for Simple Multi-Agent Systems and Static Platforms

The complexity of world modeling architectures is proportional to complexity of autonomous systems. For example, swarm members are technically limited to elementary motion and cooperation activities. Thus, the world model contains information sufficient for a few reactive responses, like finding bounding boxes of environment entities, planning a path or tracking a target. Another example is a stationary platform, which is bound to stay at the same location. Since only reactive behavior in a restricted world of interest is required, the world modeling system is rudimentary. Moreover, matching of world model elements to real entities is straightforward due to artificial functioning conditions and thorough calibration. Such world model usually contains geometry information of entities under consideration and of the proximity environment required for the task fulfillment and for safety considerations.

### World Modeling for Mobile Platforms

Mobile platforms have to be robust and maneuverable, performing a limited set of complex activities such as tracking in dense clutter and autonomous exploration. Modern world modeling systems are often oriented to SLAM, path analysis, entities tracking and collision avoidance. The absence of artificial conditions for functioning and pre-calibration lead to extensive employment of data association and fusion algorithms.

One of the goals of the world modeling of autonomous mobile platforms is an appropriate description of the world of interest suitable for exploration, reconnaissance and patrolling. A typical example is the SLAM world modeling architecture of AMROS [Emt10]. During an exploration, the AMROS robot (Fig. 1.6b) is reconstructing its environment map, represented as an occupancy grid. The modeling describes each cell $c^{x,y}$ with coordinates $x, y$ with an occupancy probability $p(c^{x,y}) = o_k^{x,y} / i_k^{x,y}$, where $i_k^{x,y}$ denotes number of inspections of the cell before the time moment $k$ and $o_k^{x,y}$ number of times that the cell has been found occupied. Hereby, the self-localization and cartography are affected by sensor uncertainties, e.g. by errors in odometry estimation due to wheel slipping. The SLAM random variables (e.g. path $\vec{\boldsymbol{x}}$, control input $\vec{\boldsymbol{u}}$, measurement $\vec{\boldsymbol{z}}$) and dependencies between the state variables are modeled as a Dynamic Bayes Network, following concepts in [Thr05], and updated by Bayesian Filtering (i.e. Extended Kalman Filter). The SLAM algorithm estimates posterior probability distribution $p(\vec{\boldsymbol{x}}^k, \Theta | \vec{\boldsymbol{z}}^k, \vec{\boldsymbol{u}}^k, n^k)$, where index $k$ denotes all elements up to time moment $k$, $\Theta$ is a set of $N$ landmarks and $n$ represents associations between observations and landmarks. The path estimate is managed by a particle filter (FastSLAM), with each particle representing a hypothesis for the true path of the robot.

Another example is given in [Mat08] and [Mat10] for intelligent sensors/actuators control. In these works, spatio-temporal probabilistic information about targets is stored in a dynamic transient world model and updated by Bayesian filtering. The interest $\Theta$ in the target is estimated by a conditional probability $P(\Theta_s^k | F_s^k(\Theta_s^{k-1}) A_k(s))$, where time-dependent functions $F$ and $A$ assign weights to the target $s$ according to the spatial proximity and goal-dependent attentional salience respectively.

### World Modeling for Intelligent Multi-Agent Systems

Usually, modeling architectures for multi-agent systems are complex and sophisticated. Such systems are designed for tasks that imply situation awareness and decision making. For example, the configuration of each sensor (e.g. drone position and orientation) has to be constantly optimized for the information flow enrichment taking into account possible configurations of other sensors. Additionally, multi-modal information fusion with non-zero network latency has to be performed real-time as well as sophisticated inferences has to be made for a situation assessment.

Since intelligent multi-agent systems imply cooperation of many agents, the world model can be shared among different devices and allow team behavior. As discussed in [Vla01], shared world models in a multi-agent system have to be mutually consistent between agents, support team modeling and coordination. A formalism for such description can be a multi-agent Markov decision process [Bou96] extended for team roles [Spa02] and team members mutual behavior modeling [Kok02]. Such formalism is presented in [Vla01] as follows: a multiagent Markov decision process with roles is defined as a tuple $< S, N, M, \{A_m\}_{m \in M}, F, Pr, R >$, where $S$ denotes a set of world states, $N$ a set of agents, $M$ a set of roles, $A_m$ a set of actions $\{a_m\}$ associated with role $m$, $F : S \times N \mapsto M$ is a role assignment function (each agent receives only one role), $Pr : S \times A_{F(s,1)} \times \ldots \times A_{F(s,n)} \mapsto [0,1]$ is a transition function that defines probability of resulting state $s_{k+1}$ after executing action $a_{F(s_t,i)}$ in the state $s_k$ by $i \in N$, and $R$ is a reward function for reaching state $s$. In this definition, a role $m$ specifies the desired behavior of the agent by defining the set of possible actions $A_m$ and its policy. Thus, each agent can perform planning and decision making taking into account its role and the current world state updated by rare synchronizations and affected by cooperation with teammates. This mutual modeling of teammates' behavior is vital, for example, when the present world state is not fully observable by some agents. Hereby, each agent has a probability distribution for the state at hand for e.g. predicting next action of another agent according to known policy.

An example of a practical realization of a shared world modeling is discussed in [Vla01] and [Gro01] as follows: a shared world modeling gathers and delivers information about the environment taking into account clutter and communication network malfunctions. The set of modeling entities includes static elements (e.g. topology) and moving entities (e.g. teammate robots and other robots). Each robot infers its own state (position, orientation, and velocity) from observations of static elements using a Kalman or particle filter.

The difference between shared and single-robot world modeling is in the fact that each robot observes other teammate robots, as well as areas or entities occluded for single robots. This information is exchanged between robots of the team and fused into their own world models. Particularly, in the case of RoboCup robot soccer competition, each robot estimates a combined state $\vec{\boldsymbol{S}}^k = [\vec{\boldsymbol{X}}^k, \vec{\boldsymbol{O}}^k, \vec{\boldsymbol{B}}^k]$, where $\vec{\boldsymbol{X}}^k = \{\vec{\boldsymbol{x}}_i^k\}, i = 1,\ldots,4$ denotes state (composed by position, orientation, velocity) of four teammate robots at time step $k$, $\vec{\boldsymbol{O}}^k$ is the state of opponents, and $\vec{\boldsymbol{B}}^k$ is the state of the ball (position and velocity). Hereby, the modeling implies two known stochastic models: a state transition model $p(\vec{\boldsymbol{S}}^k|\vec{\boldsymbol{S}}^{k-1})$ that describes kinematics of the robots and the ball (prior knowledge about system parameters combined with odometry data), and an observation model $p(\vec{\boldsymbol{Y}}^k|\vec{\boldsymbol{S}}^k)$, where $\vec{\boldsymbol{Y}}^k = \{\vec{\boldsymbol{y}}_i^k\}$ denotes collaborative observation of teammate robots (processed CCD camera images). According to Bayes' rule, the state estimate is derived as

$$p(\vec{\boldsymbol{S}}^k|\vec{\boldsymbol{Y}}^k) = \alpha p(\vec{\boldsymbol{Y}}^k|\vec{\boldsymbol{S}}^k) \int p(\vec{\boldsymbol{S}}^k|\vec{\boldsymbol{S}}^{k-1}) p(\vec{\boldsymbol{S}}^{k-1}|\vec{\boldsymbol{Y}}^{k-1}) d\vec{\boldsymbol{S}}^{k-1}, \quad (1.1)$$

where $\alpha$ is a normalizing constant.

Since observation vector $\vec{\boldsymbol{y}}_i^k = [\vec{\boldsymbol{y}}_i^k(e), \vec{\boldsymbol{y}}_i^k(\vec{\boldsymbol{x}}, \vec{\boldsymbol{o}}, \vec{\boldsymbol{b}})]$ is a composition of observed static environment and observed moving entities, the self-localization is computed by

$$p(\vec{x}_i^k|\vec{y}_i^k(e)) \propto p(\vec{y}_i^k(e)|\vec{x}_i^k) \int p(\vec{x}_i^k|\vec{x}_i^{k-1}) p(\vec{x}_i^{k-1}|\vec{y}_i^{k-1}) d\vec{x}_i^{k-1}. \tag{1.2}$$

If $\vec{y}_j^k(\vec{x}_i)$ (an observation from teammate robot $j$ of the robot $i$) is available, the self-localization estimate can be improved by treating the state estimate in Eq. 1.2 as prior

$$\begin{aligned} p(\vec{x}_i^k|\vec{y}_j^k(\vec{x}_i)) &\propto p(\vec{y}_j^k(\vec{x}_i)|\vec{x}_i^k) p(\vec{x}_i^k|\vec{y}_i^k(e)) \\ &= p(\vec{x}_i^k|\vec{y}_i^k(e)) \int p(\vec{y}_j^k(\vec{x}_i)|\vec{x}_j^k,\vec{x}_i^k) p(\vec{x}_j^k) d\vec{x}_j^k. \end{aligned} \tag{1.3}$$

The team work requires robust and efficient network communication between teammates. Due to network malfunctions and delays, a lag filtering is applied. To this end, each robot maintains a history of state posteriors that are corrected (history revision) upon receiving of an observation from teammates.

Similar approach for a team world modeling is given in [Göh07]. Hereby, the state estimate (topology and geometry information) is described within a Hidden Markov Model with information fusion performed by a particle filter.

### World Modeling for Humanoid Robots

Since the main research goal in the field of androids is to develop robots most similar to human beings, the intended capabilities imply most advanced cognition, reasoning and planning. The capabilities and the performance of the world modeling subsystem have to be similar to those of human memory. Such a system is currently not feasible, thus, researchers limit the world modeling to some realistic scenarios, e.g. fetching objects from shelves, going upstairs, loading dishwashers or playing football as team.

There are few research attempts dedicated to complex multi-purpose world modeling. For example, [Hsi03] and [Roy04] have presented a world modeling system within a 3D simulation. The simulation architecture is similar to the Open Dynamic Engine [ODE06a], which is a rigid body dynamics framework for world modeling in terms of spatial information (including shapes), colors, masses and forces. The simulation itself allows for complex modeling sufficient for non-trivial scene analysis (e.g. interpreting human commands as "bring me a red cup from the right"). Unfortunately, this simulation allows only fixed scenarios with pre-defined scenes.

A system for combined modeling and recognizing of entities is presented in [Kri13]. In this work, authors have discussed an active exploration for scenes of partially known entities and exploration approaches for autonomous modeling of unknown areas. However, the proposed approach is limited to the spatial information only.

The research group of Prof. Beetz at Bremen University has tried to broaden possible application fields for humanoid robotics. To this end, they have focused on generalizing the world modeling subsystem for performing "everyday activities". A complete discussion of meaning of everyday activities in relation to robotics is given in [And95] (this includes e.g. rational bounds for information absorption and control over environment). The research group in Bremen has narrowed the analysis of everyday activities to those that satisfy the following restrictions [Win13]:

1. complex tasks that are both common and mundane to the agent performing them;
2. those about which an agent has a great deal of knowledge, which comes as a result of the activities being common;
3. those at which adequate or satisfactory performance rather than expert or optimal performance is required.

They have created a modeling system called *CRAM*$_{\text{(m)}}$ (Cognitive Robot Abstract Machine), described in [Bee10] and [Win13]. In particular, CRAM employs *KnowRob* [Ten09], which is a knowledge processing system, particularly designed for autonomous personal robots, providing knowledge required for taking decisions. The paper [Bee10] describes KnowRob as "a first-order knowledge representation based on description logics and provides specific mechanisms and tools for action-centric representation, for the automated acquisition of grounded concepts through observation and experience, for reasoning about and managing uncertainty, and for fast inference – knowledge processing features that are particularly necessary for autonomous robot control." However, due to its logic-oriented nature, KnowRob can be classified as a task-dedicated framework for mobile robotics and is not suitable for an abstract world description.

JDL Data Fusion Model

In 1985 the U.S. Joint Directors of Laboratories (JDL) Data Fusion Group developed a data fusion model for categorizing data fusion-related functionality. This model represents a formal architecture for data fusion, world modeling and cognition systems. According to this model, the domain of *information sources* (radars, databases, etc.) serves as the input, and the domain of *human-computer interfaces* (HCI) as the output connected to the *data fusion* domain. The latter contains information processing levels. In 1998, Steinberg et al. presented an extended variant of the JDL model [Ste98] that is depicted in the Figure 1.9. In this article, the authors distinguish between the following processing levels:

- level 0 (sub-object data assessment) – estimation and prediction of signals/representative states on the basis of "raw" sensory data;
- level 1 (object assessment) – estimation and prediction of representative states on the basis of pre-processed information and dynamic model contents;
- level 2 (situation assessment) – estimation and prediction of relations among entities;
- level 3 (impact assessment) – estimation and prediction of effects on situations of planned or estimated/predicted actions by the participants, in order to include interactions and action plans of agent groups;
- level 4 (process refinement) – adaptive data acquisition and processing for supporting mission objectives.

A detailed discussion of this formal modeling description is given in [Lig09].

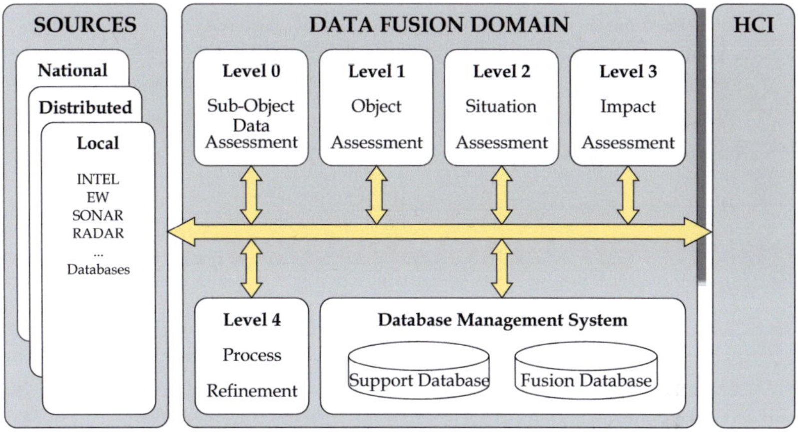


**Figure 1.9:** The JDL data fusion model [Ste98].

## Modeling of Relations

Many researchers attempt to model limited environments, thus, the developed models are suitable for a narrow set of tasks. Typically, such models contain only one or two semantic networks, such as geometrical and functional primitives [Riv95] or spatial hierarchies with additional attributes (color, weight) [Fri98], [Rie97], [Rog03]. Some researchers have analyzed relations in a broader meaning, such as contextual relations and relations presented as situations. For example, some ideas of modeling situations are presented in [Fis11]. The situation assessment can be based on the Hidden Markov Model [MD09], but this approach is dependent on training samples. [Gli06] has discussed Markov random fields employed to model contextual relationships and maximum a posteriori labeling to infer intentions.

## Prior Knowledge Employment Examples

We distinguish several prior knowledge types:

1. *intrinsic prior knowledge* – methods for: sensory information processing, information association and fusion, evolution models and inference processing. Additionally, it contains a *scheme of attributes*, describing known types of attributes (precision, coordinate system with reference, color space, etc.);
2. *topology* – an environment map;
3. *concepts* – known entities (specific exemplars or common types), constraints, etc.

*Intrinsic Prior Knowledge* Any intelligent autonomous system employs the first mentioned type – the intrinsic prior knowledge. The intrinsic prior knowledge declares modeling attributes: what attributes are under consideration, what is the information representations form and units, etc. It gives a basis for information processing and management, as well as semantics for involved attribute descriptions. For example, [Rus10] suggests formal description of values with *unit functions* that take a number as argument, e.g.:

$$Length(L_1) = Inches(1.5) = Centimeters(3.81). \tag{1.4}$$

*Topology* Topological data have to be modeled when e.g. the environment is known a priori or when the system performs SLAM. For example, all patrolling robots, many static platforms and humanoid robots model topology data required for operational and safety information processing. Many multi-agents systems, such as surveillance systems or football robotic teams, contain thorough topology description that enables the situation awareness. For example, [Göh08] and [Göh09] propose a constraints based world modeling for robot navigation within an environment of known topology (i.e. football field). A set of formal rules reduces the space of possible configurations. Additionally, the inconsistency and the ambiguity of data are quantified for constraints processing.

For intelligent mobile platforms, a pre-defined topology information "includes all of information which is available in advance, before the autonomous vehicle starts its journey. This includes for example a planned travel path, coordinates of intersections and roundabouts, and/or other relevant information about the road infrastructure, such as the number of traffic lanes" [Fur10]. This information can be presented in a formal format, like Route Network Definition File (RNDF). The persistent knowledge about the topology of the world of interest can be created during the autonomous system operation. Different algorithms, commonly named simultaneous localization and mapping (SLAM), are dedicated to this thematic, e.g. refer to [Dis01], [Dav04].

*Concepts* A modeling of concepts allows for classification and re-recognition of entities, as well as reducing configuration space with pre-defined rules and constraints. However, due to theoretical and computational complexity, researchers focus the analysis and development on a limited set of functionality. Most of the state of the art entity recognition methods are based either on sensor data (so-called global or local methods) or on models.

The global sensor data methods rely on prior knowledge containing a complete data-set describing an entity under consideration, e.g. appearance of an object from all possible directions or its 3D model [Kas13]. This data-set is compared to actual observations by e.g. an analysis of sizes and color histograms [Mac10a] or employing a Viola-Jones detector [Vio01].

The local methods represent an entity as a set of local features defined a priori that have to be matched during the recognition process (as discussed e.g. in [Shi94], edges detector in [Har88]). These local features are represented by feature descriptors, within e.g. Scale Invariant Feature Transform (SIFT) [Low04] or Speeded Up Robust Feature (SURF) [Bay08] frameworks.

The model-based recognition employs geometrical, functional and other models, allowing methods such as e.g. Pose from Orthography and Scaling with ITerations (POSIT) [Dem95] or other that are introduced in this thesis.

An overview of sensor data-based or model-based recognition (and hereby relevant representation of the prior knowledge) in application specific cases is given e.g. in [Kas13]. In the following chapters we discuss general cases for information modeling and entities recognition over arbitrary attributes.

Since we are dealing with complex concepts possibly in open world set-up, it is necessary to structure the available information. A necessity of hierarchical concepts representation e.g. in form of an ontology (refer to

Section 3.4.3), as well as philosophical discussion and practical approaches overview of a general-purpose ontologies are provided in [Rus10]. Hereby, the authors have mentioned four development ways: ontologies created by ontologists/logicians [Len90], by importing information from existing databases (e.g. Wikipedia) [Biz09], by parsing text documents [Ban08], by enticing people to enter commonsense knowledge (e.g. within Web 2.0 concept) [Sin02],[Chk05]. Also, [Rus10] argues about the employment of classes of objects (i.e. categories) as follows:

> The organization of objects into *categories* is a vital part of knowledge representation. Although interaction with the world takes place at the level of individual objects, much reasoning takes place at the level of categories. For example, a shopper would normally have the goal of buying a basketball, rather than a particular basketball such as $BB_9$. Categories also serve to make predictions about objects once they are classified.

The need for ontology modeling is discussed in detail in [Gan02]. In particular, the suggested goal of ontologies is to define concepts vital for knowledge representation in a given context. Also semantics must be associated with defined concepts, e.g. by specifying axioms.

A connection of prior knowledge (in form of an ontology) to dynamic models is discussed e.g. for deterministic categorization in [Men05] and [Go05]. To this end, the ontology is conveniently specified with the Ontology Web Language (OWL). The application of ontologies in the field of spatial and topology representation, as well as navigation is presented in [Rem98], [KB05] and [Bat04].

The prior knowledge can also include constraints and rules for concepts and elements of the dynamic model. Such constraints can be described in the ontology or in the intrinsic prior knowledge.

### Alternative Approaches

As an alternative to the considered object-oriented models, there exist other solutions such as e.g. neural networks, logic-based models, etc. They also have their drawbacks and applicability issues, e.g. neural networks represent black-box models trained over known entities and scenarios and are more suited for sensor data-based classification (e.g. refer to [Jam01]). As another example, logic-based systems process information in terms of rules and predicates according to relations within data. This approach describes well actions and situations [McC63] but fails to compete with model-based systems in processing sensor data. As a live example, we can mention again the KnowRob [Ten09] – the knowledge processing framework targeted for mobile and humanoid robotics (e.g. used in RoboEarth as a local knowledge base). It provides tools for information acquisition, representation and reasoning and is designed to serve as a common semantic framework for integrating multimodal information. It employs description logic as a formalism for the information representation. Description logic is a family of formal knowledge representation languages, similar to first-order logic (refer to e.g. the OWL). A good overview of graphical and logic networks is given e.g. in [Jai12]. As we have mentioned earlier, this approach perfectly models actions and situations but is not suitable for abstract modeling.

## 1.4 Thesis Information

### 1.4.1 Motivation

We have mentioned many existing works dedicated to world modeling – starting from low-level data storages with Bayesian filtering, SLAM and neuronal networks to Bayesian networks, complex entities modeling and situation analysis. Now we are going to discuss motivation points of our own work.

A comprehensive and real-time world modeling for intelligent autonomous systems raises problems that are too challenging for current theories and technologies (the extensive discussion of challenges in modern intelligent robotics and existing systems addressing them are presented previously in the Section 1.3.3). Thus, most of existing works are concentrated on a narrow set of questions with convenient assumptions. There are very few analyses dedicated to the creation of a universal object-oriented platform capable of storing, managing and analyzing arbitrary information pieces required for an arbitrary intelligent autonomous system. This includes dynamic modeling of entities under open-world assumption, connection of an abstract dynamic information to prior knowledge and bringing semantics into models. Moreover, there are almost no modeling systems employing methods for both qualitative and quantitative information analysis, vital for proactive behavior. Also, there are too few works for information sufficiency analysis relative to given tasks. Moreover, there is a lack of physical limits implications involved in modeling, e.g. the uncertainties are considered usually only while establishing the covariance formulation. Finally, many of existing theoretical analyses are dedicated to common theoretical problems and are proved with artificially constructed abstract examples. In our work we have tried to overcome these limitations.

Scope

In the following sections, we will discuss a general platform for modeling an abstract information in an object-oriented way and generalize the information handling. On its basis, we will consider a generalized dynamic model connected to prior knowledge database. Moreover, we will consider numerous methods for probabilistic information analysis within such object-oriented framework. Hereby, we will examine a model-based classification. Due to arising tasks complexity, we will limit our research to high-level modeling (entities and their relations) and leave a low-level data pre-processing and SLAM out of scope. Also, we will deliberately avoid Bayesian networks and semantic analysis of scenes discussions. In the next section we will describe the delivered solutions for the ARMAR and NEST platforms and the published research works.

### 1.4.2 Contributions

Within the scope of this work, we have attained a row of theoretical and experimental results in the field of world modeling as well as gathered significant experience and expertise. The covered topics include concepts and approaches for dynamic and prior knowledge modeling, information association, fusion and management as well as their practical realization and experimental evaluation.

One of notable topics we have researched and developed is a dynamic modeling of arbitrary entities (even those that are unknown or cannot be classified). In this model, the description of entity's attributes is given in form of marginal or joint degree-of-belief distributions (including mixed joints of discrete and continuous distributions), which allowed a unified probabilistic information representation empowered by Bayesian fusion.

Also, we have researched and implemented various assessment methods over the abstract models. Since model elements – called representatives – are interconnected within the dynamic model as multiple semantic nets, we have employed a complex analysis of entity relations, e.g. qualitative situation and context assessment. The introduced entropy calculus enables numerical estimates of the information sufficiency (relative to given tasks or to maximum possible content) of representatives, their groups and the whole model given sensor parameters. This enables the extensive information deficit analysis.

Additionally, we have introduced a certain nomenclature for the world modeling, which defines clearly and consistently all the crucial parts of the modeling domain and which had not existed in such form before.

Next, we have researched possibilities of prior knowledge modeling and employment of persistent information in dynamic models. Since many challenges for intelligent autonomous systems involve not only the abstract modeling of the environment but also semantics of its elements, as well as enrichment of the observed information with domain knowledge, it is crucial to be able to model the pre-defined expert knowledge and to connect it to the dynamic model. To this end, we have developed an ontology framework with multiple hierarchies and mechanisms for matching of ontology elements to dynamic model representatives on the basis of structural (e.g. Tanimoto metric) and value (e.g. Kullback-Leibler divergence) similarities. This enables a probabilistic classification with sub-sequent weighted update with prior knowledge.

Finally, we have systematically analyzed possibilities for physical parameters implication. Many existing researches were dedicated to abstract theoretical studies with experimental tests over artificial tasks and simulations. In contrary, we have examined the effects of limits imposed on physical parameters (either due to finite sensor accuracy or due to the semantics of the task). Here we have introduced the concept of a least discernible quantum to entropy calculus unification as well as to resampling procedure. As a generalization of this idea, we have developed a generalized functions framework for quantification of the accuracy of arbitrary distributions and demonstrated its use with the practical problem of reducing the number of components in Gaussian mixtures and optimal sampling for particle filtering. Also, we have mentioned advantages of generalized functions applied to matching of dynamic model elements to prior knowledge.

In addition to theoretical analysis, we have implemented and experimentally evaluated the proposed methods and approaches, including creation of dynamic modeling and prior knowledge modeling subsystems. For the information interchange between these two subsystems, we have implemented mechanisms for probabilistic information association (e.g. for the connection of observations to dynamic model elements and the latter to prior knowledge) and Bayesian fusion (e.g. for tracking and dynamic model extension with prior knowledge). In order to deploy these subsystems and mechanisms on real robotic systems, we have implemented a distributed cross-platform infrastructure with declared interfaces for interactions with other subsystems of a robot. Finally, we have implemented a visualization subsystem for attributes monitoring and 3D-visualization of the model contents.

At the experimental evaluation phase, we have conducted several practical tests on two already mentioned robotic platforms: the ARMAR-III robot and the NEST system and analyzed obtained results.

A part of research and development results was presented in national and international scientific conferences, as well as in several robotics books: [Küh10], [Bel10], [Bau10a], [Bau10b], [Ghe10], [Bel11], [Bel12a], [Bel12b], [Bel12c], [Bel12d], [Pak13].

### 1.4.3 Structure of this Thesis

We discuss our work in five chapters, each presenting a separate topic. In Chapter 1 we provide an introduction into autonomous systems and world modeling domains and overview of the state of the art in the field of interest. Also, we describe the analysis motivation, context and goals, as well as point out the achieved research and technology contributions.

In Chapter 2, we discuss in detail the world modeling requirements and concepts, including probabilistic information representation in form of degree-of-belief distributions. In addition, we define here a comprehensive nomenclature relevant to world modeling. Further, we introduce one of the most powerful approaches in the statistical mathematics – the Bayesian Framework, followed by filtering techniques such as Kalman filter. We complete the information fusion discussion with such topics as Gaussian mixture reduction and particle filtering.

In Chapter 3, we present advanced techniques in dynamic world modeling, namely distinguishing of levels of abstraction, modeling of arbitrary entities with Progressive Mapping, performing qualitative and quantitative information assessment. Additionally, we describe prior knowledge modeling in form of attribute schemes and ontologies of known concepts. Next, we introduce connection mechanisms of prior knowledge to dynamic models in form of probabilistic classification and concepts learning.

Further, we discuss experiments and give evaluation of different aspects of the world modeling in the Chapter 4. We start the discussion with the modeling system realization, continue with test scenarios and demonstrators and finish with detailed examples and obtained results.

The final Chapter 5 contains a summary of research and technology contributions and presents an outlook.

# 2

# World Modeling and Information Fusion

## 2.1 Introduction

This chapter introduces basic world modeling concepts and techniques that allow creation of simple world models. Widely used in static and mobile platforms and swarms, such world models are pretty straightforward: they contain mainly spatial and relational information of the environment. This information allows to perform targets tracking, navigation/SLAM, path and collision avoidance analysis. One of the most popular approaches here is the object-oriented world modeling with continual update from sensors. In short, world of interest entities are coupled with virtual representatives – which is natural and intuitively understandable compared to other approaches, e.g. neuronal networks. In the following, we are going to discuss such object-oriented modeling as well as model updating mechanisms (e.g. with newly acquired information).

### 2.1.1 Information Acquisition

Modern autonomous systems constantly acquire information about the surrounding environment in order to reactively or proactively respond to the occurring changes. The information acquisition starts with the observation of the environment made by a sensor. Usually, there are many sensors of different types installed within the autonomous system, e.g. radar, LIDAR, (stereo-)cameras, microphones and so on. This *multimodality* allows for broader range of available information, as well as cross-confirmation and complementation.

The raw data stream from sensors is processed by low-level filters implemented at hardware (e.g. FPGA) or software levels in order to reduce the data flow and to extract important features or parameters. At this point, data processing (such as image segmentation or 3D-spacing) is performed at high rates.

### 2.1.2 Information Association and Fusion

The acquired features and values have to be combined with the already known information. This means that the incoming information has to be matched to world model elements, e.g. an observed person has to be associated to its representative in the model. In practice, information association is quite challenging due to sensor limitations, clutter, partial visibility and other issues. For example, it is hard to track a person (constantly trace his or her position) in a dense crowd.

As soon as the incoming data are associated with the world model elements, it is necessary to fuse the information into the existing description. The information fusion is also challenging due to uncertainties and complex relations between descriptions. In the simplest case, the known information can be overwritten with an incoming information piece. In this case, however, the clutter can affect the estimation of the current state dramatically. Therefore, it is advantageous to consider existing information along with the incoming information for the current state estimation. To this end, we can take the existing description and propagate it to the observation moment according to some evolution model which allegedly describes how the system changes with time. For example, if we see a falling ball, we can propagate its coordinates estimate according to physical laws of motion. This prediction – prior information about the system state at the given time moment – is updated then with incoming sensory information, giving us the posterior state estimate. In this chapter we will consider such recursive two-step prediction-update fusion resulting in estimation of posterior states. These estimates are descriptions with uncertainty of the world of interest elements, which is, in fact, our up-to-date dynamic world model. Before advancing to information management topics, it is necessary to define basic information description notions.

### 2.1.3 Nomenclature and Mathematical Notions

We denote vectors by small bold letters with vector head, e.g. $\vec{\boldsymbol{a}}$ and $\vec{\boldsymbol{\mu}}$, and matrices by bold capitalized letters, e.g. $\boldsymbol{\Sigma}$ and $\boldsymbol{\Delta}$. From now on, we use only column-vectors, e.g.:

$$\vec{\boldsymbol{x}} = \begin{bmatrix} x \\ y \end{bmatrix}.$$

For the notions of probability theory we use the nomenclature similar to those employed in [Das08]. Let some attribute be modeled as a random variable, denoted as $A$. A specific value that $A$ may assume we denote as e.g. $a$ (scalars or facts) or $\vec{\boldsymbol{a}}$ (vectors). The probability distribution of a random variable $A$ is a function of $A$ that specifies probabilities for each possible value of $A$. Throughout this thesis we consider only normalized probability distributions, which we call as *probability distribution functions* (pdf). A pdf of a discrete random variable we call as *discrete pdf* or *probability mass function* (p.m.f.). On the other hand, a pdf of a continuous random variable is denoted as *continuous pdf* or *probability density function* (p.d.f.).

Let us first examine the discrete case. Here, we have a random variable $A$ with possible values $a_1, \dots, a_N$. The p.m.f. $p$ of $A$ is denoted as $p(a_i), i = 1, \dots, N$ or in short $p(a)$. The expected value or *expectation* of $A$ is defined as

$$\mu \equiv E[A] = \sum_i a_i p(A = a_i). \tag{2.1}$$

The *joint p.m.f.* of two discrete random variables $A$ and $B$ is denoted as $p(A,B)$ that specifies probabilities for each possible ordered pair $(a,b)$, where $a$ and $b$ are possible values of $A$ and $B$ respectively. The probability for each given pair is given as $p(a_i,b_j) = p(A = a_i, B = b_j)$. This joint probability can be marginalized by summing up over one variable: $p(A) = \sum_i p(A,B_i)$.

When we consider a probability of an event depending upon another event, we employ a *conditional probability.* In this case, we place sign | (meaning "given") between random variables: $p(A|B)$ (probability of $A$ given $B$). We can express conditional probability through joint probability as

$$p(A|B) = \frac{p(A,B)}{p(B)}. \tag{2.2}$$

It is easy to extend our formulas to continuous and vector cases. The expectation for a p.d.f. over $A$ is given by

$$\vec{\boldsymbol{\mu}} \equiv E[A] = \int_{-\infty}^{\infty} \vec{\boldsymbol{a}} p(\vec{\boldsymbol{a}}) d\vec{\boldsymbol{a}}. \tag{2.3}$$

Other important notions are variance $\mathrm{var}(A)$ and standard deviation $\sigma$:

$$\begin{aligned} \mathrm{var}(A) &= \sigma^2 \\ &= E[(A - E[A])^2] \\ &= \int_{-\infty}^{\infty} (a - E[A])^2 p(a) da. \end{aligned} \tag{2.4}$$

Similarly, we may define the covariance $\mathrm{cov}(A_1, A_2)$ of two attributes as:

$$\begin{aligned} \mathrm{cov}(A_1, A_2) &= \sigma^2_{A_1 A_2} \\ &= E[(A_1 - E[A_1])(A_2 - E[A_2])] \\ &= \int_{-\infty}^{\infty} \int_{-\infty}^{\infty} (a_1 - E[A_1])(a_2 - E[A_2]) p(a_1, a_2) da_1 da_2. \end{aligned} \tag{2.5}$$

The deviations of $A_1$ and $A_2$ constitute a covariance matrix $\boldsymbol{\Sigma}_{A_1 A_2}$. We construct the covariance matrix by pair covariances of vector elements of both attributes (a kind of multi-dimensional variance):

$$\boldsymbol{\Sigma}_{A_1 A_2} = \begin{bmatrix} \sigma^2_{A_1} & \sigma^2_{A_1 A_2} \\ \sigma^2_{A_2 A_1} & \sigma^2_{A_2} \end{bmatrix}. \tag{2.6}$$

### Gaussian Distributions and Mixtures

Since numerous parameters of real systems can be represented with sufficient accuracy by Gaussian distributions (refer to central limit theorem) or Gaussian mixtures (refer to [Pla00], [Fel71] and [Sor71]), we have to introduce them into our mathematical nomenclature.

A continuous random variable $A$ defined over $(-\infty, +\infty)$ range is said to follow univariate *Gaussian* or *normal distribution* if it has the p.d.f. given by:

$$p(A) = N(\mu, \sigma; a) \equiv N(\mu, \sigma) \equiv \frac{1}{\sigma\sqrt{2\pi}} \exp\left\{ -\frac{(a-\mu)^2}{2\sigma^2} \right\}. \tag{2.7}$$

Its extension to multi-dimensional case (e.g. $D$-dimensions), called multivariate (or $D$-variate) Gaussian / normal distribution, is given by the formula:

$$\begin{aligned} p(A) &= N(\vec{\boldsymbol{\mu}}, \boldsymbol{\Sigma}; \vec{\boldsymbol{a}}) \\ &\equiv N(\vec{\boldsymbol{\mu}}, \boldsymbol{\Sigma}) \equiv \frac{1}{\sqrt{(2\pi)^D det|\boldsymbol{\Sigma}|}} \exp\left\{ -\frac{(\vec{\boldsymbol{a}} - \vec{\boldsymbol{\mu}})^T \boldsymbol{\Sigma}^{-1} (\vec{\boldsymbol{a}} - \vec{\boldsymbol{\mu}})}{2} \right\}, \end{aligned} \tag{2.8}$$

where $\vec{\boldsymbol{a}}$ is a $D$-dimensional attribute value and $\boldsymbol{\Sigma}$ is a covariance matrix.

A *Gaussian mixture* (GM) is a p.d.f. formed by a weighted sum of $K$ Gaussian distributions (kernels). Hereby, we assume each kernel to be $D$-variate Gaussian distribution. The GM is then given by:

$$GM\left( \{ w_k, \vec{\boldsymbol{\mu}}_k, \boldsymbol{\Sigma}_k \}_{k=1}^K ; \vec{\boldsymbol{a}} \right) = \sum_{k=1}^{K} w_k N(\vec{\boldsymbol{\mu}}_k, \boldsymbol{\Sigma}_k; \vec{\boldsymbol{a}}), \tag{2.9}$$

with $w_k$ specifying weights.

As we have already mentioned, numerous parameters of real systems can be represented with sufficient accuracy by Gaussian distributions or Gaussian mixtures. Thus, it is convenient to approximate our distributions with relatively handy Gaussians or GMs. To this end, there are many criteria and algorithms for selecting optimal mixture parameters, discussed e.g. in our analysis [Pak13], as well as in [Num83], [Zee97], [Pla00].

## 2.2 World Modeling

### 2.2.1 Components

A posterior state estimate of an entity (i.e. representative) defines a hypothesis about the entity. A collection of all actual representatives gives information about the dynamic state of the world of interest and thus is called *dynamic model*. This information along with prior knowledge about the environment composes the world model. Here, under prior knowledge we usually understand persistent information about known entities and classes of entities. In [Bel12d], we have stated: "When preparing a system for operation, it is possible to equip the system with definitions of all the relevant concepts that are supposed to be encountered within its operational tasks. Since these concepts are defined prior to operation, they constitute a priori knowledge." As already mentioned in Chapter 1, there is intrinsic prior knowledge that describes methods for sensory information processing, information association and fusion, evolution models and inference processing. Moreover, intrinsic prior knowledge contains directly or indirectly global scheme of attributes, describing known types of attributes (precision, coordinate system with reference point, color space, etc.). A detailed discussion on attributes scheme is presented in [Bel12d].

### 2.2.2 Requirements

The world modeling system has to comply with several requirements, in order to be useful in practice [Bel12d]:

- *correctness* – modeling of dynamic information as well as prior knowledge has to describe the world of interest in a sufficiently correct (relative to designed tasks) way;
- *minimality* – keeping the world model as small as possible, in order to cope with combinatorial explosion of the processing time (e.g. during data association or entities classification);
- *universality* – being able to model all required types of information – probability distributions, relations, symbolic facts, etc. Moreover, being able to handle unknown entities and extending concepts under open world assumption (a dynamic and changing world of interest, which consists of entities of unforeseen types). Preferably, the world modeling has to handle uncertainties in a universal way for all elements;
- *semanticity* – conforming acquired information to prior knowledge (our believes about the surrounding environment) and enriching it with semantics;
- *robustness and efficiency* – storing information in a robust and efficient way (e.g. for real-time operation);
- *dispatch* – acquiring information from other modules of the autonomous system and providing relevant information back (i.e. serving as an information hub);
- *clarity* – the structure and contents of the world model has to be transparent and clear, interpretable for both humans and machines. In this analysis the structure is assumed to be object-oriented.

The contents of the world model has to describe information with uncertainties. Thus, it is necessary to define the information representation – the exact form of descriptions that are acquired and stored in the world model.

### 2.2.3 Information Representation

#### Generalization Hierarchy

Recording of an entity description implies a representation of its attributes. In ideal case, these attributes can be represented as a set of precise values (e.g. a numerical value 117 or a fact value CAT). However, such representation does not include uncertainties, emerging during the information acquisition process. Moreover, this often leads to momentary shifts in modeling values with each measurement and to overlapping of these values (e.g. spatial overlapping of representatives in the world model). A generalization of the information representation was described in [Ghe08] and our previous works [Bel10], [Bel12d] as follows: at simplest, we can describe uncertainties with error values, which allow more complete representation of measured values. The next level of the generalization can be introduction of relations between attribute descriptions, representatives or concepts or representation of attributes with probability distributions over possible values, e.g. by means of *degree-of-belief* (DoB), as shown in Fig. 2.1. DoB p.d.f. representation allows for probabilistic description without necessity of subject's statistics (i.e. prior observations) and enables Bayesian treatment for all attributes (represented by a universal data type). In Fig. 2.2 two attributes of a quantitative continuous (p.d.f.) and a qualitative discrete (p.m.f.) types are represented as DoBs.

A cross-correlations between attributes can be incorporated into DoB distributions by introducing of pair joint distributions (e.g. joint DoB distribution of the position and type) as shown in Fig. 2.2. In this case, it is possible to combine discrete and continuous distributions into one common multidimensional frame. Further, it is possible to combine DoB distributions into arbitrary group joint distributions. At the end of the generalization, there is a joint DoB distribution of all attributes. Such description is still computationally too hard for intelligent autonomous systems within real environments. An overview of the generalization levels is presented in Fig. 2.3.

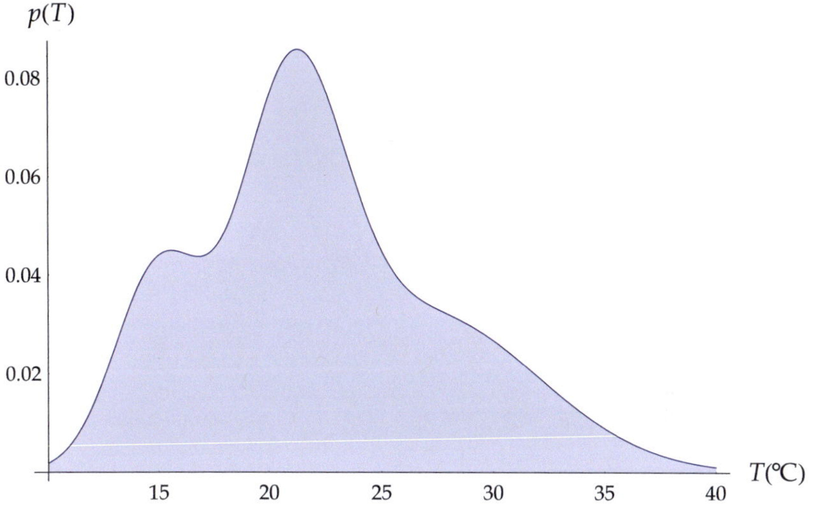


**Figure 2.1:** A typical degree-of-belief distribution for the (TEMPERATURE attribute).

## 2.2.4 Relations

Modeling of relations, mentioned in Fig. 2.3, implies modeling connections among attribute descriptions, representatives and concepts, e.g. as a semantic network (Fig. 2.4). In common case, intelligent autonomous systems require a variety of relations (e.g. "part of", "is a"), leading to multiple semantic networks over the same elements (Fig. 2.5) [Bel10].

## 2.2.5 Information Flow

In [Bel12d] we have described an information flow in intelligent autonomous systems (Fig. 2.6) as follows: the world of interest consists of entities, which an autonomous system observes with sensors. This results in "raw" sensory data (e.g. video stream or acoustic signals). The sensor data are analyzed on the basis of prior knowledge (intrinsic and concepts) and are fused into the dynamic model. As some information is unobservable (e.g. sit-

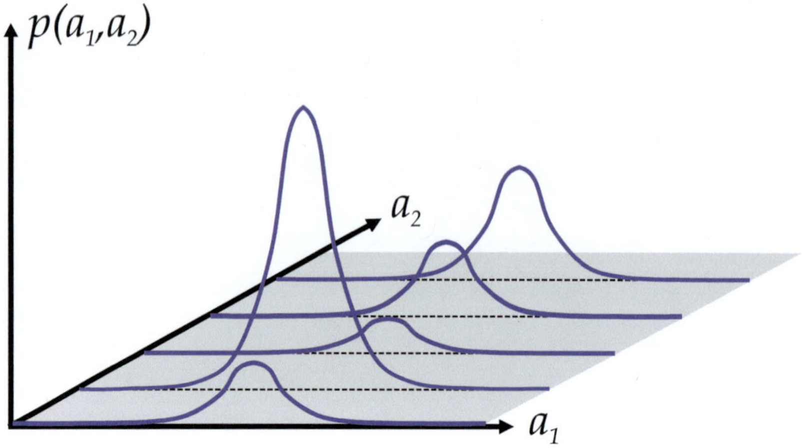


**Figure 2.2:** Joint degree-of-belief of two attributes of different types: $a_1$ is quantitative continuous (e.g. POSITION), $a_2$ is qualitative discrete (e.g. TYPE).

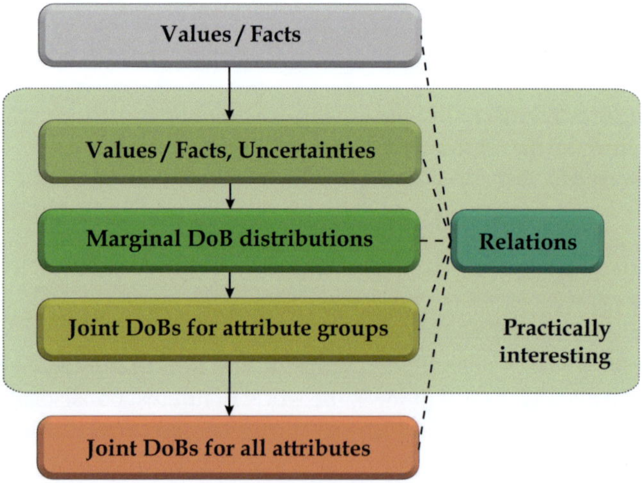


**Figure 2.3:** Generalization hierarchy of the information representation [Bel12d].

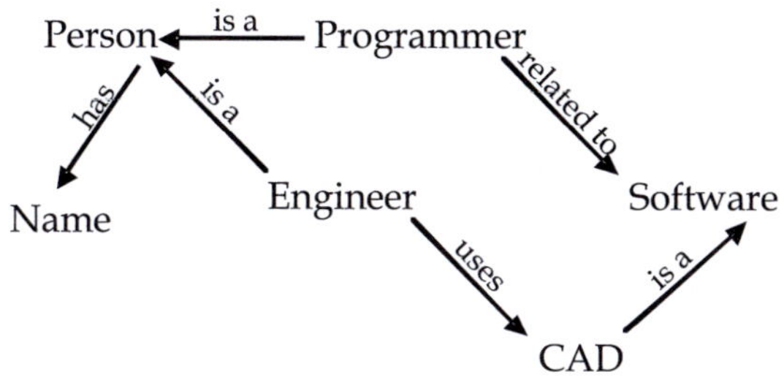


**Figure 2.4:** An example of a semantic network [Bel10].

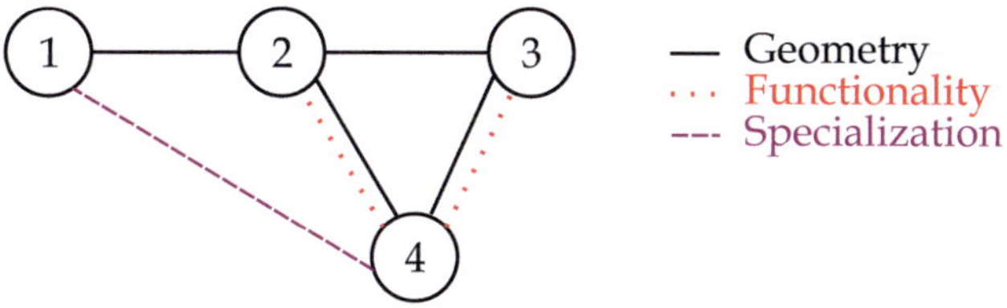


**Figure 2.5:** Multiple semantic networks over same elements [Bel10].

uations, context, hidden entities), cognitive processes reason about present context and make inferences by means of prior knowledge. The resulting information is passed to dynamic model and other subsystems (e.g. planned actions to actuators that interact with the world of interest).

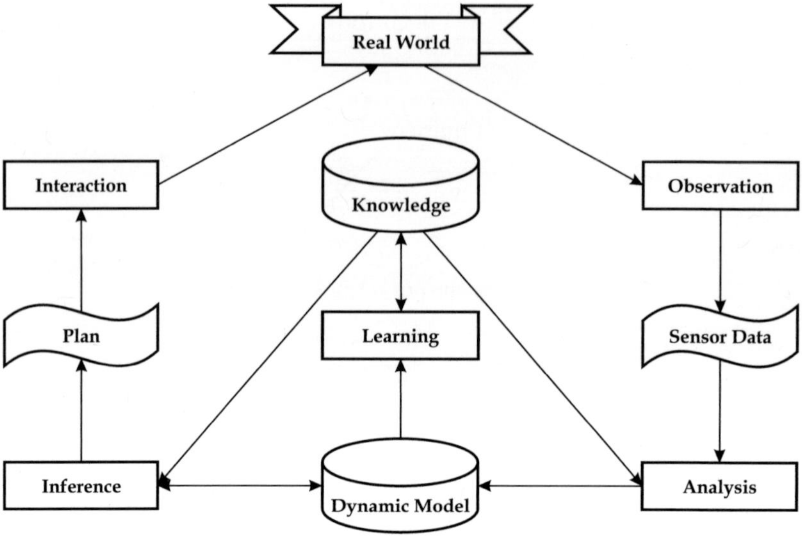


**Figure 2.6:** Information flow: information (waved boxes), processes (boxes), databases (cylinders) and input/output (arrows) [Bel12d].

## 2.3 Information Fusion

The contents of the world model are updated with newly acquired information. The sensory data is considered to be "low-level, raw" data that need to be processed. For example, a video stream can be analyzed for 3D segmentation or finding entities, gestures can be recognized with pixel flow processing and colors corrected with light balancing. This pre-processing occurs in a perception subsystem. The results of such pre-processing are fused into existing descriptions in the world model according to some fusion process, which was defined e.g. in [Whi91] as:

> A process dealing with the association, correlation, and combination of data and information from single and multiple sources to achieve refined position and identity estimates, and complete

> and timely assessments of situations and threats as well as their significance.

or later in [Ste98] as:

> Data fusion is the process of combining data to refine state estimates and predictions.

For a more detailed introduction into fusion subjects, refer to e.g. [Das08] and [Mac10b].

Below we formalize the ideas of the information acquisition and fusion within a Bayesian Framework and explain them on examples, presenting several corresponding state of the art methods.

### 2.3.1 Bayesian Framework

One of the most powerful and well-developed paradigms in modern statistical theory is the Bayesian theory. This theory is often applied to problems encountered in information association and fusion due to several reasons: first of all, the Bayesian theorem (Eq. 2.11) operates with beliefs affected by uncertainties. This naturally matches the fundamental notion "degree-of-belief" (DoB), employed throughout current analysis. Second, the recursive form of Bayesian inference is convenient for updating the world model with sequential observations coming from sensors. Moreover, the Bayesian fusion is a handy approach for processing observations coming from multiple heterogeneous sensors and dealing with clutter. Finally, the Bayesian framework is very popular within many areas of computer science and especially robotics, forming quasi state of the art framework in numerous application fields. Before advancing to Bayesian fusion methods, it is necessary to clearly define a formalism for the state representation and discuss the underlying Bayes' theorem.

#### State Estimate

The parameters of real entities are observed by sensors and presented by attribute descriptions assigned to corresponding representatives. Such attribute descriptions are DoB distributions, e.g. in form of marginal or joint

Gaussian mixtures. A set of attribute descriptions forms the state estimate, which is a probabilistic statement about the attribute values of an entity given all available information. Here we have to make three simplifications, which will help us to deal with exponential explosion in information processing time in the case of realistic number of parameters (e.g. describing all system parameters with all dependencies under the open world assumption):

1. we consider only marginal or pair joint probability distributions for the representation of attribute descriptions;
2. we consider any state estimate evolution with some time quantum $\Delta t$;
3. we assume all state estimates to possess *Markov property*, namely, a conditional state pdf depends only upon the previous state and not all past states:

$$p(\vec{\boldsymbol{s}}_k|\vec{\boldsymbol{s}}_{k-1}, \vec{\boldsymbol{s}}_{k-2}, \ldots, \vec{\boldsymbol{s}}_{k-n}) = p(\vec{\boldsymbol{s}}_k|\vec{\boldsymbol{s}}_{k-1}), \tag{2.10}$$

   where $\vec{\boldsymbol{s}}_k$ denotes a multidimensional state random variable at time step $k$. In order to take information uncertainty into account, we represent our knowledge about the state with a pdf $p(\vec{\boldsymbol{s}})$.

We model a representative with index $i$ at time step $k$ as a DoB distribution $p(\vec{\boldsymbol{s}}_k^i) \equiv p(e_k^i, \vec{\boldsymbol{a}}_k^i)$, where $e_k^i$ is a binary random variable specifying the representative's existence and $\vec{\boldsymbol{a}}_k^i := \left[a_k^{i,1}, \ldots, a_k^{i,n_a}\right]^T$ is a vector with $n_a$ discrete (e.g. TYPE) and continuous (e.g. X-coordinate) attributes. From now on, we assume no cross-correlations between attributes of different representatives.

### Bayesian Theorem

The Bayesian theory is defined by the Bayesian theorem, which expresses a *posterior probability* through a *prior probability* and a *likelihood function*:

$$p(\vec{\boldsymbol{h}}|\vec{\boldsymbol{e}}) = \frac{p(\vec{\boldsymbol{e}}|\vec{\boldsymbol{h}})\,p(\vec{\boldsymbol{h}})}{p(\vec{\boldsymbol{e}})}, \tag{2.11}$$

where $\vec{e}$ denotes *evidence* (e.g. observation), $\vec{h}$ stands for *hypothesis* (e.g. state estimate), $p(\vec{h}|\vec{e})$ is the *posterior probability* (probability of the hypothesis given the evidence), $p(\vec{e}|\vec{h})$ means the likelihood, $p(\vec{h})$ stands for *prior probability* (hypothesis information before the evidence was introduced), and $p(\vec{e})$ is the *marginal likelihood.*

In the simplest case, we interpret the Eq. 2.11 as follows: a new pdf of our estimate given new observation is proportional to the likelihood for this observation times pdf of the previous state estimate. For convenience, we often assume the likelihood to be a Gaussian distribution with mean equal to observation and standard deviation to sensor uncertainty. The main problem of most tracking and information fusion tasks – i.e. how to update the state estimate given new sensory information – is solved by Eq. 2.11 resulting in $p(\vec{h}|\vec{e})$ (state estimate pdf given an observation).

More explicit introduction to the Bayesian approach is given in numerous books, e.g. in [Cha11]. A detailed discussion of the Bayesian formalism for information representation and fusion is given in our analysis [Bau10a], which we omit in the current work.

### Creation and Aging

During the information acquisition we can meet a situation that new information pieces do not match with any available representative in the model. It can be clutter or signals coming from a new, previously unobserved entity. In the simplest case, we can create a corresponding new representative (i.e. a hypothesis about possible entity) in the dynamic model. This operation changes the whole world state estimate and represents a simple version of *Multiple Hypothesis Tracking* (MHT). In principle, we could always create a new representative upon receiving an unassignable observation, though, this will lead to creation of "ghost" representatives due to clutter or missmatched observations coming from already known entities. A creation threshold can be set, for example, to the calculated posterior probability that a new entity is detected and the detected entity exists (the existence attribute has to exceed some threshold $p(e_k^i = 1) > D_c$, see Fig. 2.8). In practical realization, it is convenient to create a virtual representative and accumulate further observations presumably related to it. After several reconfirmations, the existence probability of this virtual representative goes above some threshold $D_c$. This triggers switching a flag from VIRTUAL to REAL (e.g. $e^i_k = 0 \rightarrow 1$), leading to the appearance of the representative in the dynamic model.

Similarly, we remove a representative from the dynamic model if the maximum of the existence DoB falls below some deletion threshold, i.e. $p(e^i_k = 1) < D_d$. Many entities within the world of interest are not persistent, e.g. an orange on a kitchen table will probably disappear after a while because someone will take it away. Since we want to keep dynamic models slim (minimality requirement in the Section 2.2.2), all representatives that became non-relevant to the current scene has to be removed from the dynamic model. The usual practice in tracking is to remove representatives after several time steps (or video frames) of the entity absence. We consider in [Bel10] and [Küh10] an alternative approach, called *aging*, which was inspired by ideas in [Ghe08]. We assume that each representative has to be reconfirmed from time to time. In the absence of re-confirmations, its existence probability will decrease with passed time (e.g. by exponential decrease with aging factor $F_{aging}$). If it lowers below some threshold $D_d$, the representative is deleted from the dynamic model. In this case, the removed representative can be transferred to prior knowledge as "known entity" for future re-recognitions. Hereby, we consider three important issues:

- in order to inform the autonomous system that a representative is going to be deleted soon, we introduce a reconfirmation threshold $D_r$. Upon going below this threshold, a system trigger is set to indicate the need of observation of the corresponding entity (validation need). The validation itself can be performed any time while the existence value is within $D_r - D_d$ (quantized by time step $\Delta t$, which is defined by e.g. system clock frequency or time intervals between occuring observations);

- the aging factor $F_{aging}$ differs for different concepts. For example, a kitchen table is more persistent than an orange, resulting in softer decrease of the table existence probability. The concept type introduces a correction of the first order ("large" contribution) to the aging factor. The second order ("smaller" contribution) can be a context dependency, e.g. party time or siesta. We can generalize corrections in the form:

$$F_{aging} = F_0 + F_{concept} + F_{context} + \ldots + F_N. \tag{2.12}$$

- since not every representative is required on each time step $k$, we consider postponed aging as reasonable: upon a request on a specific representative in the world model, its existence probability is calculated from the last update up to time step $k$. Hereby, one has to find a tradeoff between CPU-time consume (minimal by postponed aging) and memory occupancy (minimal by existence recalculation at each $k$).

The necessity of the aging mechanism is clear: without creation there would be only an empty scene and without deletion – a model overpopulated with "ghost" representatives. Obviously, $D_c > D_d$ ensuring that a created representative has a lifetime greater than zero. This statement leads to a life-cycle hysteresis depicted in Fig. 2.7. An example of the overall existence probability lifetime within a dynamic model is presented in Fig. 2.8.

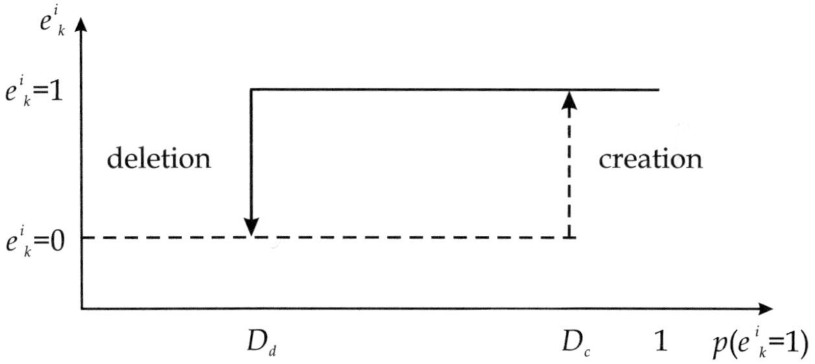


**Figure 2.7:** A hysteresis for the representative creation and deletion.

As we mention in [Bel10], a set of aging factors determines memory's temporal boundaries. Namely, it defines for how long the system usually preserves representatives. This allows for numerical estimation of memory usage for an arbitrary time step.

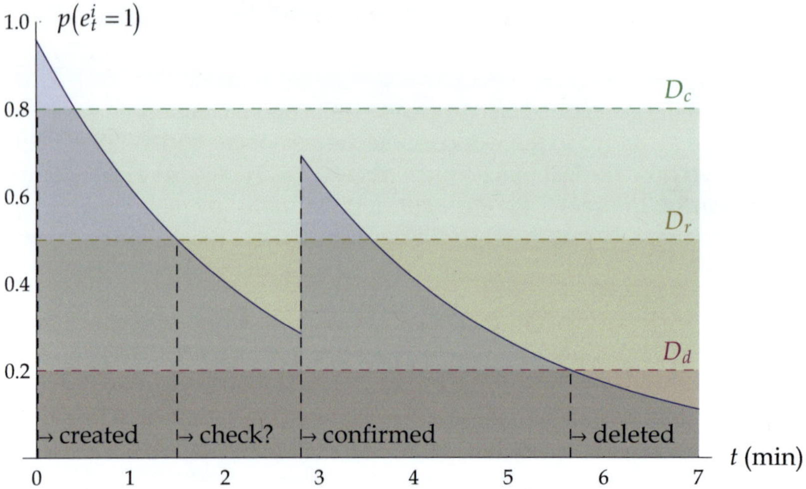


**Figure 2.8:** Existence probability lifetime within a dynamic model.

There are creation and deletion mechanisms incorporated into state of the art filtering algorithms. For example, track initiation and track termination are fully integrated into the association and smoothing parts of the integrated probabilistic data association (IPDA) algorithm [Muš94]. Another example is the filters based on Finite Set Statistics (e.g. probability hypothesis density filters) with the idea to combine all sensor measurements into a meta-observation and all possible targets into a meta-predicted target using a multiple-target transition density, a multiple-target likelihood and a reference measure [Vo06].

Similarly to the representative's aging, we define an attribute's aging [Ghe08], [Bel10]. In the case of a known evolution model, we adjust the covariance matrix of the estimate according to the evolution matrix, random process noise and passed time (refer to Kalman filtering in Sec. 2.3.2). If the evolution model is not known, we propagate the uncertainty with the random process noise only, which is dependent upon attribute type, context, etc. At last, this propagation will disperse the probability distribution, leading to flatter pdf with larger uncertainty (Figure 2.9).

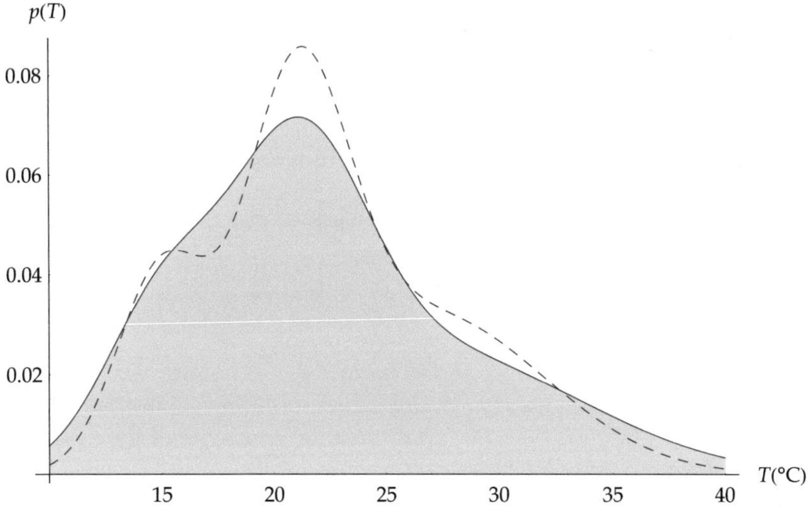


**Figure 2.9:** Attribute's aging: dashed line represents previous attribute description, solid line – aged estimate.

In the following sections, we will discuss in details several iterative filtering algorithms, which allow incremental update of the state estimate with new information (e.g. with acquired sensory information). This update mechanism along with attributes scheme (i.e. intrinsic prior knowledge) and stored states of all representatives (i.e. dynamic model) is the typical world modeling subsystem for many modern intelligent autonomous systems.

### 2.3.2 Kalman Filter

One of the well-known filtering algorithms is the *Kalman filter* (KF) [Kal60], [Gel74], [BS88], also known as *linear quadratic estimation* (LQE). The KF analyzes recursively noisy data streams from sensors and derives the optimal state estimate of the system (i.e. minimizes the squared error of the estimation). On the one hand, the recursive nature and simplicity of the filtering gives huge advantages in practical application of the algorithm. On the other hand, it assumes that system dynamic and measurement equations are linear and measurement and noise terms are Gaussian distributed, which is not always true. However, many system parameters can be described by Gaussian distributions or at least by Gaussian mixtures (acceptable for e.g. KF ensembles), which was experimentally proved acceptable for practical applications by, for example, Kalman filter usage in the Apollo navigation computers and many other systems. Thus, we consider KF as practically valuable for our information update mechanisms due to simplicity, robustness and relatively small processing costs.

### Physical Model

A physical model represents our prior knowledge about physical laws and restrictions applied to our system and the surrounding environment. The Kalman filter assumes that the (unobservable) true state of the system $\tilde{\vec{s}}$ evolves from time step $k-1$ to step $k$ as:

$$\tilde{\vec{\boldsymbol{s}}}_k = \boldsymbol{F}_k \tilde{\vec{\boldsymbol{s}}}_{k-1} + \boldsymbol{B}_k \vec{\boldsymbol{u}}_k + \vec{\boldsymbol{w}}_k, \tag{2.13}$$

where $\boldsymbol{F}_k$ is a propagation matrix (dynamic model for a projection of the state into future), $\boldsymbol{B}_k$ is a control-input model matrix, $\vec{\boldsymbol{u}}_k$ is a control vector and $\vec{\boldsymbol{w}}_k \propto N(0, \boldsymbol{Q}_k)$ is a random process noise with a covariance $\boldsymbol{Q}_k$.

Next, each observation $\vec{\boldsymbol{\mu}}$ is assumed to comply with the following model:

$$\vec{\boldsymbol{\mu}}_k = \boldsymbol{H}_k \tilde{\vec{\boldsymbol{s}}}_k + \vec{\boldsymbol{v}}_k, \tag{2.14}$$

where $\boldsymbol{H}_k$ is an observation model, and $\vec{\boldsymbol{v}}_k \propto N(0, \boldsymbol{R}_k)$ is an observation noise with a covariance $\boldsymbol{R}_k$.

### Algorithm

The Kalman filter calculates state estimate in two steps: prediction and update. The prediction step propagates the state to the next time stage taking into account the physical model, known control inputs and arising uncertainties (Fig. 2.10a). The update step combines the prediction with new information from sensors (Fig. 2.10b).

The prediction step gives the state estimate $\hat{\vec{s}}_{k|k-1}$ and its covariance matrix $\hat{\boldsymbol{P}}_{k|k-1}$ as following:

$$\hat{\vec{s}}_{k|k-1} = \boldsymbol{F}_k \hat{\vec{s}}_{k-1|k-1} + \boldsymbol{B}_k \vec{\boldsymbol{u}}_{k-1}, \tag{2.15}$$

$$\hat{\boldsymbol{P}}_{k|k-1} = \boldsymbol{F}_k \hat{\boldsymbol{P}}_{k-1|k-1} \boldsymbol{F}_k^T + \boldsymbol{Q}_{k-1}. \tag{2.16}$$

The update step combines the prediction with new observation:

$$\hat{\vec{s}}_{k|k} = \hat{\vec{s}}_{k|k-1} + \boldsymbol{K}_k \left( \vec{\boldsymbol{\mu}}_k - \boldsymbol{H}_k \hat{\vec{s}}_{k|k-1} \right), \tag{2.17}$$

$$\hat{\boldsymbol{P}}_{k|k} = (\boldsymbol{I} - \boldsymbol{K}_k \boldsymbol{H}_k) \hat{\boldsymbol{P}}_{k|k-1}, \tag{2.18}$$

where $\boldsymbol{I}$ is the identity matrix and $\boldsymbol{K}_k = \hat{\boldsymbol{P}}_{k|k-1} \boldsymbol{H}_k^T \boldsymbol{S}_k^{-1}$ is the Kalman gain with innovation $\boldsymbol{S}_k = \boldsymbol{H}_k \hat{\boldsymbol{P}}_{k|k-1} \boldsymbol{H}_k^T + \boldsymbol{R}_k$.

In Sec. 4.4.2 we present an example application of the Kalman filter.

### 2.3.3 Kalman Filter Extensions

#### Nonlinear

The Kalman filter assumes that system dynamic and measurement equations are linear (e.g. the current state is a linear function of the previous state and observations are linear functions of the state). If these conditions are no longer satisfied, KF is no more the optimal estimator. Many successful attempts have been performed to adapt KF to nonlinear processes, e.g. [BS01], [Leo02], [Arr98]. Most notable approaches here are Extended Kalman filter (EKF) and Unscented Kalman filter (UKF) (discussed e.g. in [Göh09] and [Cha11]). In this analysis, we omit discussion of nonlinear KF variants and later introduce a particle filter as an alternative approach (see Section 2.3.5).

#### Non-Gaussian

The Kalman filter works with Gaussian distributions (e.g. representation of a state estimate, prediction and likelihood). Sometimes the estimate of a real system has to be described by a more complex distribution, then the

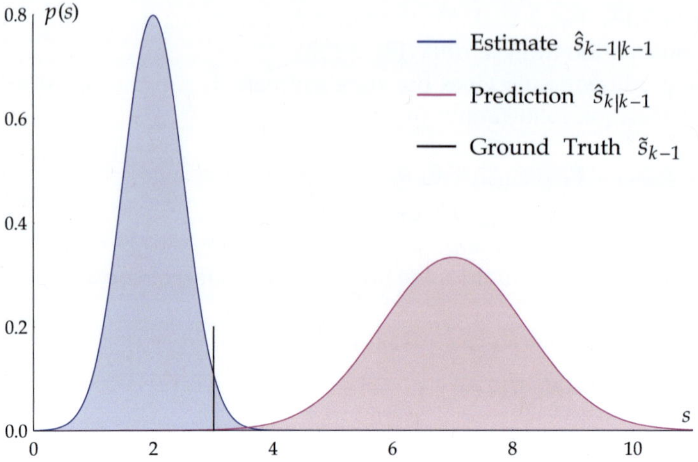


**(a)** The prediction step propagates a state estimate with an increase of variances/covariances (due to random process noise);

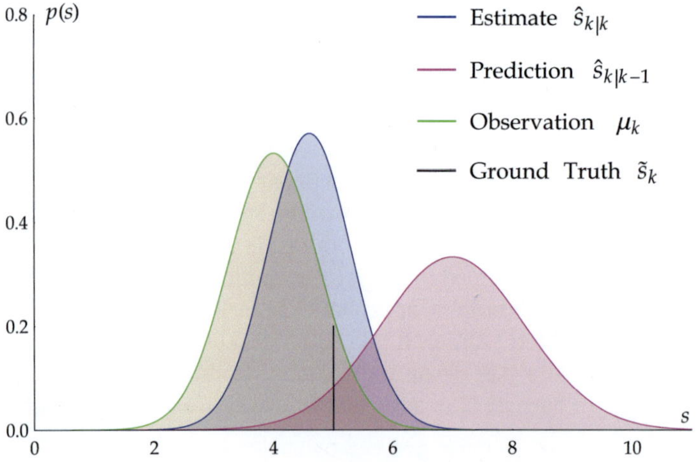


**(b)** The update step combines an observation with the prediction. Here we see a decrease of the variances/covariances due to compacter form of the likelihood;

**Figure 2.10:** 1D Kalman filter algorithm: prediction and update steps.

Gaussian mixture representation could be advantageous. In some cases, tracking can be performed in closed form with inputs and outputs at each step as GMs (e.g. [Rei79], [Pao94]). In general, the state estimate can be represented by a GM with kernels representing our hypothesis. A discussion of this approach, called Multiple Hypotheses Tracking, can be found e.g. in [Göh09]. Hereby, a GM pruning and merging has to be performed, keeping the number of mixture components "small", thus, preventing possible explosion in complexity [Aus00]. In this analysis, we omit discussion of GM tracking approaches (though present our method for GM pruning in Section 2.3.4). Instead, we introduce a particle filter with a least discernible quantum (LDQ) conversion to Gaussian mixture, required e.g. for visualization and resampling, in Section 2.3.5.

#### Data Association with Extended Attributes Set

Classical tracking approaches focus on kinematic parameters of the system: position, velocity, acceleration. However, in the case of targets moving side by side or crossing their paths the data association (i.e. matching observations to representatives) becomes a challenging task. This ambiguity cannot be fully resolved even by advanced observation-to-representative association as in joint probabilistic data association. If we consider tracking of additional information, e.g. color and size, such situations become much easier to handle, as it is discussed in the Section 4.4.2. Moreover, additional modeling attributes extend the dynamic description and improve classification as we show it in Section 4.4.5.

#### Multiple Targets and Observations

The Kalman filter works with one target entity. In practice, many tasks (e.g. traffic control, collision avoidance, etc.) require simultaneous tracking of several targets. In this case, multiple observations are coming from sensors matching different entities. We need to assign the incoming observations to representatives in order to perform correct information update. This so-called data association (or better information association, since we are working on the high-level modeling) is one of the main topics in multi-target multi-sensor tracking [Hal04], [Lig09]. In principle, we can use the *nearest neighbor algorithm* [RL96] for the observation assignment. Here, if the distance between observed values and values of known representatives is larger than a certain threshold, this observation is considered unassigned (which leads to possible creation of a new representative). More general approach is addressed in *probabilistic data association* (PDA) [BS09] and its extension for creation and deletion of targets – the *integrated probabilistic data association* (IPDA) algorithm [Muš94]. For the case of weighted assignment of multiple observations to multiple targets, there is further extension of KF, called joint integrated probabilistic data association (JIPDA), presented in [Muš04], [Göh09]. A general Bayesian formalism for multi-target tracking with target existence is given in e.g. [Ver05] and [Hor09], as well as in our previous work [Bau10a].

### 2.3.4 Weak Distance Between Distributions

#### Introduction

In [Pak13] we discuss several problems of modern tracking algorithms and propose corresponding solutions. State of the art tracking techniques depend on information representation (i.e. multi-modal distributions in multi-dimensional spaces) and management. These distributions (e.g. probability density functions) are often given by Gaussian distributions or Gaussian mixtures. Such information representation is usually convenient for analysis and visualization. Moreover, in some cases tracking can be performed in closed form (e.g. Kalman filter for Gaussian distributions and algorithms discussed in [Rei79], [Pao94], [Göh09] with inputs and outputs at each time iteration as GMs).

However, there are cases in which distributions have to be converted to other representations. For example, a particle filter in [Kot03] employs posterior distributions represented by GMs with the prediction step requiring sampled distributions (linear combinations of Dirac delta-functions). Moreover, by tracking with pure Gaussian mixtures, the number of kernels usually grows after each time iteration, and we are forced to prune the mixtures or find GMs similar to the original distributions but with fewer components in order to prevent the explosion in complexity [Aus00]. The distribution conversions can include changing information representation to histograms or moments (e.g. norm, mean, variance, etc.). The corresponding quality losses have historically been assessed ad-hoc with help of different methods [Wil03a], [Run07], [Cro11]: information-theoretical concepts (e.g. Shannon entropy), functional metrics (e.g. integrated square difference), or heuristics. The problem of resulting pruning criteria and similarity metrics is the requirement of abstract universality. Their performance and correctness is typically evaluated on abstract artificial examples. In [Pak13] we propose a more theoretically sound similarity metric and apply it to the problem of GM reduction as follows.

New challenges in tracking and GM analysis are naturally leading us to extension of our scope that is yet limited to regular functions, which map arguments to values as $y = f(x)$. So, we are going to introduce so-called *generalized functions* [Gel64] to our statistical framework. The generalized functions are determined by generalized moments, namely, by the following integrals:

$$\langle f, \psi \rangle = \int f(\vec{\boldsymbol{x}}) \cdot \psi(\vec{\boldsymbol{x}}) d\vec{\boldsymbol{x}}, \tag{2.19}$$

where $f(\vec{\boldsymbol{x}})$ is a *functional* applied over some so-called *probe function* $\psi(\vec{\boldsymbol{x}}) : \mathbb{R}^n \to \mathbb{R}$ of some class $T$. This generalization allows employment of broader class of functions, e.g. singular Dirac delta-function, but requires proper selection of class $T$. In order to understand Dirac delta-function and the above integral, we illustrate them with a historical reference [Dob14]:

> Paul Dirac introduced in 1926 his celebrated $\delta$-function via the relation
>
> $$u(x) = \int \delta(t - x) u(t) dt, \tag{2.20}$$
>
> where $\delta(x) = 0$ if $x \neq 0$. Such a "function" is zero everywhere except at the origin, where it becomes infinite in such way as to ensure
>
> $$\int_{-\infty}^{\infty} \delta(x) dx = 1. \tag{2.21}$$

> ...
>
> We cannot see elementary particles like electrons, but we can observe the point where the electron strikes the screen. To describe this phenomenon mathematically, Dirac suggested using integration of two functions, one of which corresponds to a particle, and the other one, called the "probe" function, corresponds to the environment (as a screen). Hence, the $\delta$-function operates on the "probe" functions according to Eq. 2.20. The delta-function can be interpreted as the limit of a physical quantity that has a very large magnitude for a very short time, keeping their product finite (i.e. the strength of the pulse remains constant).

The defining of the corresponding class $T$ appears to be an advantage: indeed, additionally to the constraints usually imposed on $T$ (like compactness of support, integrability, etc.), probe functions bring into play physical characteristics of our system, i.e. by encoding the uncertainty of measurements and the precision required by our tasks. For example, let us consider tracking of a person: all micrometer-size features are meaningless (a person has no POSITION definition at that scale), irrelevant (nobody requires such precision) and almost impossible (no conventional sensor delivers such accuracy). Hereby, probe functions can introduce so-called *configuration scale.* The importance of this scale is repeatedly underlined in all real tasks: e.g. the sensor pitch and distance to target impose different precision scales upon $x$, $y$ and $z$ coordinates (*sensor accuracy*), as well as ten-meter resolution can be insufficient for indoor tracking but suitable for tracking a person in a forest (*task relevance*). Moreover, a metric for differences between distributions has to deal with arbitrary distributions localized beyond the sensor range. As we have stated in [Pak13], from this point of view the process of GM reduction and finding of a difference metric between GMs created for abstract cases is non-optimal for real tasks. The introduction of artificial metrics and analysis of moments (e.g. refer to [Wil03a] and [Cro11] for an overview) has little meaning without employment of corresponding physical parameters, which can be described by probe functions.

An assessment of probability distributions discrepancy is performed by state of the art methods like, for example, Kullback-Leibler (KL) divergence

[Run07], [Bel12c] or integrated squared difference (ISD) [Wil03a], [Wil03b], [Hub08]. Such functional difference metrics restrict the type of functions they allow. For example, ISD in form of

$$d_{ISD}(f,g) = \int \left(f(\vec{x}) - g(\vec{x})\right)^2 d\vec{x} \tag{2.22}$$

is useful only for relatively smooth functions, whose values $f(\vec{x})$ and $g(\vec{x})$ are defined everywhere [Pak13]. Also, both metrics are inappropriate for sampled distributions (i.e. linear combinations of Dirac delta-functions). We will return to these metrics in 3.5.2 for a state of the art approaches analysis. For now, we are going to discuss our alternative approach for comparing generalized distributions.

### Weak Difference Metric

*Definition* In order to overcome mentioned limitations associated to state of the art difference metrics, we can switch to a comparison of distributions based on their generalized moments. For this, we combine Eq. 2.19 and Eq.2.22 as follows. At first, we select from the infinitely-dimensional "continuous" space of probe functions a set $T$ of those probe functions that are interesting for our goals (i.e. reflect system limitations and task requirements). Next, we rewrite the Eq.2.22 as the difference of given functionals operating over selected probe functions (generalized integrated squared difference, GISD):

$$d_T^*[f,g] \triangleq \int_T w[\psi] \left(\langle f,\psi\rangle - \langle g,\psi\rangle\right)^2 D\psi, \tag{2.23}$$

where $D$ is some measure in $T$ and $w[\psi] \geq 0$ is a real-valued functional denoting the density of probe functions per unit volume of $T$. Hereby, $w[\psi]$ serves as a weighting functional over $\psi$, performing "fine-tuning" of the difference estimation, since $T$ can include all probe functions suitable for our goals in principle.

The Eq. 2.23 is as general as possible and, unfortunately, hardly applicable in practice. In order to obtain a practically interesting equation, we limit the infinitely-dimensional space of probe functions by some parametrization. Here, we restrict probe functions to a subset of parametric functions: $\psi(\vec{\boldsymbol{x}}) \rightarrow \psi(\vec{\boldsymbol{x}}, P) \equiv \psi_P(\vec{\boldsymbol{x}})$, where $P$ denotes finite-dimensional set of parameters of some $\Omega$ region. The corresponding set $T$ is formally written as $T = \{\psi_P | P \in \Omega\}$. Given some probe functions $\psi_P(\vec{\boldsymbol{x}})$, we can rewrite Eq. 2.23 as:

$$d_T[f,g] \triangleq \int\limits_{P\in\Omega} w(P) \left\langle f-g,\psi_P \right\rangle^2 dP, \tag{2.24}$$

where $w(P) \geq 0$ is the density of probe functions per unit volume of the parameter space.

Since we are working with generalized functions, the Eq. 2.24 can find a distance between e.g. a Gaussian mixture $\sum_{k=1}^{K} m_k N\left(\vec{\boldsymbol{\mu}}_k, \boldsymbol{\Sigma}_k; \vec{\boldsymbol{x}}\right)$ and a sampled distribution $\sum_{l=1}^{L} n_l \delta(\vec{\boldsymbol{x}} - \vec{\boldsymbol{x}}_l)$.

*Discussion* In order to illustrate the suggested formulas, let us consider a simple example of tracking with a stationary camera. It is quite often the case that camera processing modules deliver a variety of attributes, like 3D coordinates, size and color information of tracked entities. Thus, the corresponding set $T$ is formed by $\psi_P(\vec{\boldsymbol{x}}, \vec{\boldsymbol{c}}, \vec{\boldsymbol{s}})$ probe functions, where $\vec{\boldsymbol{x}}$ denotes coordinates, $\vec{\boldsymbol{c}}$ – color and $\vec{\boldsymbol{s}}$ – size. For simplicity, let us consider only the position information, i.e. $\psi_P(\vec{\boldsymbol{x}})$.

In order to employ probe functions, we need to select an appropriate to our task concrete form (e.g. flat or Gaussian) and choose the set of parameters $P$ (parametrization). Here we have numerous possibilities, for example:

1. continuous – we associate each space point to a probe function. Each function is, for example, represented by a uniform distribution within a sphere with the center at $\vec{\boldsymbol{\mu}}$ and radius $r$ as $\psi_{\{\vec{\boldsymbol{\mu}},r\}}(\vec{\boldsymbol{x}})$. Alternatively, the form can be an ellipsoid or the function can be a Gaussian function, i.e. parametrization with mean and covariance matrix $\psi_{\{\vec{\boldsymbol{\mu}},\boldsymbol{\Sigma}\}}(\vec{\boldsymbol{x}})$. For convenience, we will refer to probe functions by $\vec{\boldsymbol{\mu}}$, which will serve as a kind of label or marker: $\psi_{\vec{\boldsymbol{\mu}}}$;

2. discrete – we partition the space into a grid, where each node contains one probe function (e.g. uniform or Gaussian function).

The corresponding schematic is represented in Fig. 2.11.

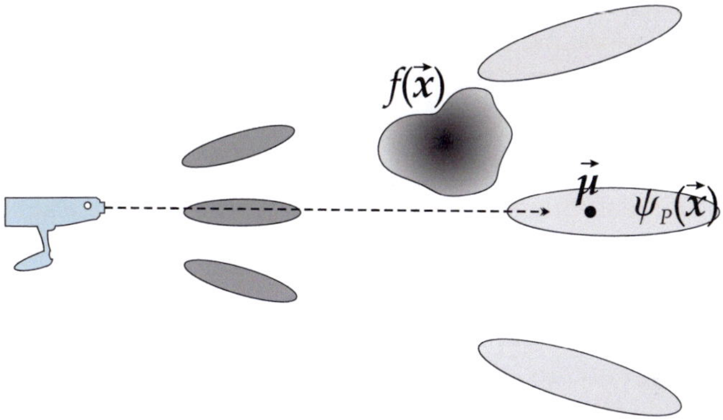


**Figure 2.11:** A camera tracks an entity, which position is estimated with the distribution $f(\vec{x})$. A single observation delivers an entity position that matches the probe function $\psi_P(\vec{x})$ at $\vec{\mu}$. Ellipses represent the primary localization areas of p.d.f.s.

It is interesting, how to select $T$ with probe functions appropriate to our task. Previously, we have discussed a configuration scale with relation to sensor accuracy and task relevance. In our example, we have to encode the camera uncertainty and the model uncertainty (associated with the relation of $\vec{x}$ and $\vec{\mu}$) by probe functions. It is convenient to take for this purpose likelihood functions $p(\vec{\mu}|\vec{x})$ with $\int p(\vec{\mu}|\vec{x})d\vec{x} = 1$, which incorporate sensor uncertainty. This sensor uncertainty specifies the maximum possible configuration scale limit (resolution), while the task relevance defines the minimum allowed (sufficient) scale limit.

Moreover, it is possible that differences at some regions of the parameter space has to be "weighted" more than in the other regions. Thus, we might need some weight function, further referred as *utility function* $\Omega_\psi(\vec{\mu})$. So, we define the set $T$ as a set of probe functions represented as a product of the utility and likelihood functions:

$$T = \left\{\psi_{\vec{\mu}} | \vec{\mu} \in \Omega \subset \mathbb{R}^n, \psi_{\vec{\mu}}(\vec{x}) = \Omega_\psi(\vec{\mu}) \cdot p(\vec{\mu}|\vec{x})\right\}, \tag{2.25}$$

where the components has the following meaning and traits:

- the utility function $\Omega_\psi(\vec{\boldsymbol{\mu}})$ introduces the relative utility assigned to a particular measurement. For example, this utility function can suppress regions irrelevant to our task (e.g. areas outside the tracking region). We assume it to be integrable and to possess some characteristic scale $s_\Omega$, which is determined by the given task;
- the normalized likelihood function $p(\vec{\boldsymbol{\mu}}|\vec{\boldsymbol{x}})$ is well-localized in the $\vec{\boldsymbol{x}}$-space around fixed $\vec{\boldsymbol{\mu}}$ with a characteristic scale $s_p$ (physically possible and task required resolution limit in the problem, e.g. $3\sigma$ value), where $s_p \ll s_\Omega$.

The integral $\int f(\vec{\boldsymbol{x}}) \cdot \psi_{\vec{\boldsymbol{\mu}}}(\vec{\boldsymbol{x}}) d\vec{\boldsymbol{x}}$ provides the "response strength" for each $\vec{\boldsymbol{\mu}}$.

In the simplest case, the utility function can be taken constant (i.e. all target locations are equally important). However, in [Pak13], we consider a more reasonable candidate – a Gaussian kernel function:

$$\Omega_\psi(\vec{\boldsymbol{\mu}}) = \exp\left(-\frac{\vec{\boldsymbol{\mu}}^2}{2R^2}\right), \tag{2.26}$$

which adds the following features:

1. the exponent assigns more weight to places closer to camera (*relative weighting*);
2. Gaussian distributions, which are handy to use in our approach, have infinite support. However, distributions arising in tracking usually do not grow exponentially at infinite scaling, so we identify Eq. 2.26 with a well-localized ellipsoidal window. The maximum scale $s_\Omega$, which is proportional to $R$, defines the window for the distributions comparison (we ignore tails of $f$ outside that window);

In order to understand the roles of $w(P)$ and $\Omega_\psi(\vec{\boldsymbol{\mu}})$, let us consider the particle filtering (for a brief topic overview refer to Sec. 2.3.5). Here, the function under consideration is usually sampled in one of two ways: either spatial densities of samples are proportional to the pdf (Monte Carlo sampling) and weights are uniform or positions are distributed uniformly within a well-localized support and weights are assigned according to the pdf. In this analogy, the concentration factor corresponds to $w(P)$ and the weights to $\Omega_\psi(\vec{\mu})$. In the simplest case $w(P) = 1$, we have uniformly distributed probe functions limited within a rectangular volume, e.g. as in Fig. 2.12a. After a spatial transformation $P^* = \tau(P)$, we receive a new $\Omega$ with differently distributed probe functions (e.g. Fig. 2.12b). In this case, we have the following GISD transformation:

$$d_T[f,g] = \int_{P\in\Omega} \langle f-g,\psi_P\rangle^2 \, dP, \tag{2.27}$$

$$\rightarrow d_T^*[f^*,g^*] = \int_{P^*\in\Omega} \underbrace{\left(\frac{\partial P}{\partial P^*}\right)}_{w(P^*)} \langle f^*-g^*,\psi_P^*\rangle^2 \, dP^*, \tag{2.28}$$

with $w(P^*)$ denoting the transformation Jacobian.

The weights in $w(P)$ and $\psi_P$ are interchangeable by a renormalization process. Indeed, in our parametrization, we can apply Eq. 2.26 to Eq. 2.24 and assume $w(\vec{\mu}) = 1$. Then the metric is:

$$d_T[f,g] = \int \left( \int \left(f(\vec{x}) - g(\vec{x})\right) p(\vec{\mu}|\vec{x}) d\vec{x} \right)^2 e^{-\frac{\vec{\mu}^2}{R^2}} \, d\vec{\mu}, \tag{2.29}$$

We can obtain the same expression by assuming $w(\vec{\mu}) = \Omega_\psi^2(\vec{\mu})$ and $\psi_{\vec{\mu}}(\vec{x}) = p(\vec{\mu}|\vec{x})$. The Eq. 2.29 demonstrates that there is no distinct separation between weight function $w(P)$, the norm of $\psi_P$ and limits of $\Omega$ due to homogeneity of Eq.2.24.

By weighting the difference of two distributions, we can bring many physical parameters into consideration. As we have mentioned above, we can use relative weighting to assign more weight to places closer to camera. More convincing example is the employment of a noise distribution. Indeed, by calculation of usual squared difference of two 1D distributions (Fig 2.13a), we can consider the integral $\int \left(f(x) - g(x)\right)^2 dx$. A consideration of the noise at present, described by an "exotic" function $n(x)$ (Fig. 2.13b), we can weight our difference according to physical noise as $\int \left(\frac{f(x)-g(x)}{n(x)}\right)^2 dx$.

*Alternative Probe Functions* In [Pak13] we discuss different choices of probe functions as follows:

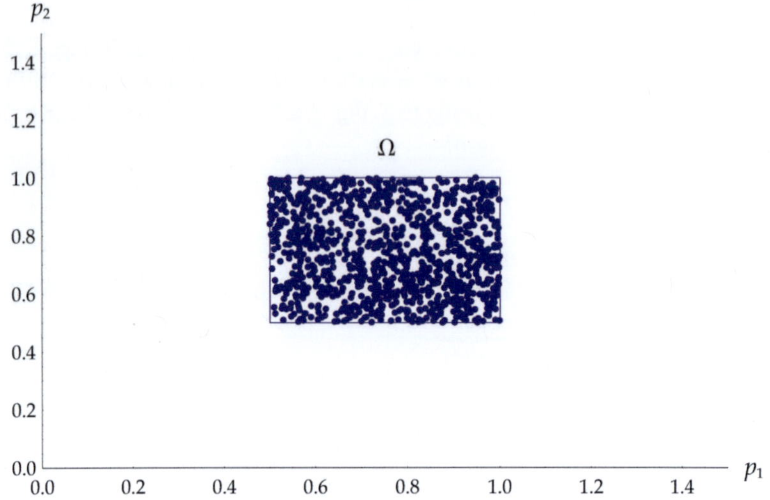


**(a)** First representation;

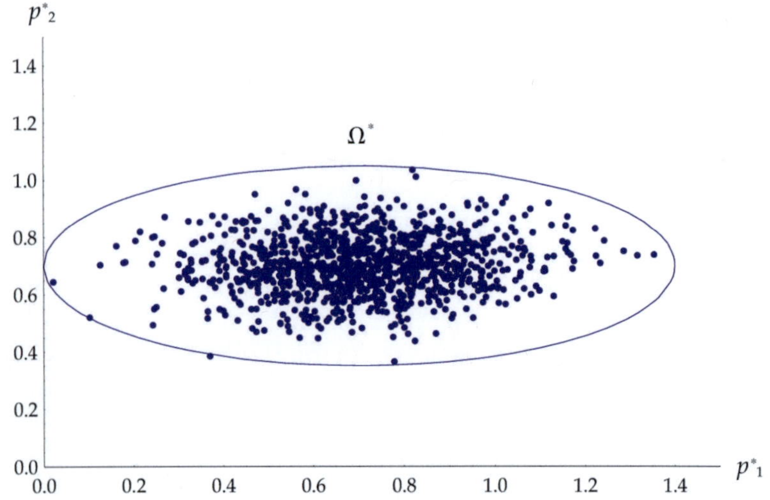


**(b)** Second representation;

**Figure 2.12:** Ω (outlined area) with probe functions (blue dots).

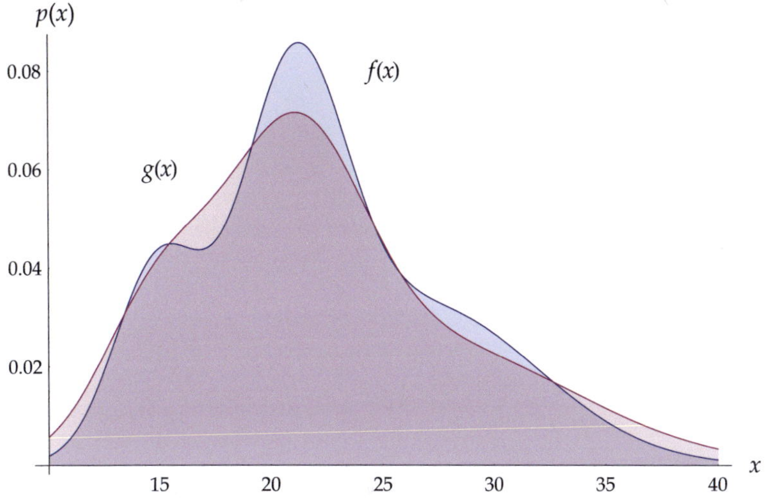


**(a)** Two distributions under comparison;

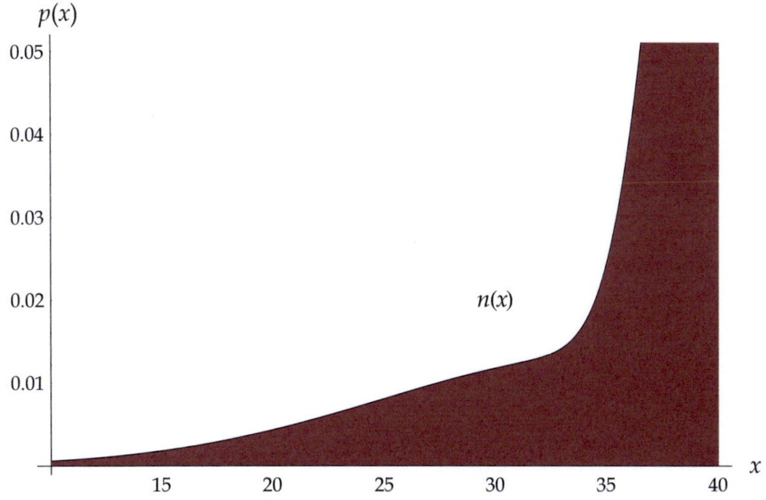


**(b)** Noise function;

**Figure 2.13:** Distributions and noise function examples.

1. *delta-functions:* under the assumption of an infinitely precise camera of unlimited range with equally important $\vec{\mu}$'s, we can assign $\psi_{\vec{\mu}}(\vec{x}) = \delta(\vec{\mu} - \vec{x})$. In this case, the probe function set is $T = \{\delta(\vec{\mu} - \vec{x}) \mid \vec{\mu} \in \mathbb{R}^n\}$ and Eq. 2.24 is equivalent to Eq. 2.22.

2. *histograms:* by partitioning a physically interesting region into $M$ rectangular areas $R_m$ and pre-integrating such sectors, we assign probe functions as *unit indicator functions* of areas corresponding to these sectors (i.e. histogram bins) as

$$T = \left\{\psi_m(\vec{x}) \mid \left\{ \begin{array}{l} \vec{x} \in R_m : \psi_m(\vec{x}) = 1 \\ \vec{x} \notin R_m : \psi_m(\vec{x}) = 0 \end{array} \right., m = 1, \ldots, M \right\}.$$

In this case, the Eq. 2.24 is the squared histograms difference:

$$d_T[f,g] = \sum_{m=1}^{M} (h_m[f] - h_m[g])^2, \tag{2.30}$$

$$h_m[\bullet] = \int_{R_m} \bullet(\vec{x}) d\vec{x},$$

where "$\bullet$" denotes either $f$ or $g$;

3. *unity:* by allowing only one probe function $T = \{\psi = 1\}$, we get the corresponding moment $|f| \triangleq \int f(\vec{x}) d\vec{x}$. The zero distance $\langle f - g, \psi \rangle^2$ delivers all $g$ normalized to $|f|$ (i.e. with equal integral: $|g| = |f|$);

4. *polynomials:* by extending singular set $T$ with further components, i.e. $T = \{\vec{1}, \vec{x}^0, \ldots, \vec{x}^n\}$, we bound all equivalent $g$'s to reproduce moments of $f$, e.g. :

$\vec{x}^0$ : norm $|g| = |f|$;

$\vec{x}^1$ : mean value $\langle \vec{x} \rangle_g = \langle \vec{x} \rangle_f$;

$\vec{x}^2$ : dispersion $\langle \vec{x}\vec{x}^T \rangle_g = \langle \vec{x}\vec{x}^T \rangle_f$;

$\ldots$;

5. *segments:* by partitioning into semi-infinite sectors $S_{\vec{\gamma}} = \{\vec{\boldsymbol{x}} | x_i < \gamma_i, i = 1,\ldots,n\}$ with corresponding probe functions as unit indicators. The metric becomes ISD between the Kolmogorov cumulative distributions for $f$ and $g$ [Han08];

6. *windows:* [Han08] has suggested finite windows $W_{\vec{\boldsymbol{a}},b} = \{\vec{\boldsymbol{x}} | b > |\vec{\boldsymbol{x}} - \vec{\boldsymbol{a}}|\}$. The corresponding unit indicator functions are considered as a superior alternative to cumulative distributions, though hardly computable;

7. *discrete set:* Since the integral in 6. can be difficult to compute in closed form (depending on $f$, $g$ and likelihood functions $p(\vec{\boldsymbol{\mu}}|\vec{\boldsymbol{x}})$), we might sample the $\Omega_\psi(\vec{\boldsymbol{\mu}})^2$ with $L$ samples, create corresponding probe functions $T = \{\psi_l \equiv p(\vec{\boldsymbol{\mu}}_l|\vec{\boldsymbol{x}}), l = 1,\ldots,L\}$ and approximate GISD with the sum:

$$d_T[f,g] = \sum_{l=1}^{L} \left(\langle f,\psi_l\rangle - \langle g,\psi_l\rangle\right)^2. \tag{2.31}$$

The corresponding metric was introduced into our framework and evaluated experimentally (refer to Sec. 4.4.2 and Sec. 4.4.3).

## 2.3.5 Particle Filter

### Introduction

In the case of nonlinear system dynamic and measurement equations and arbitrary form of probability distributions, in general, there exists no efficient way to compute posterior probability distributions due to complicated integrals and absence of closed form solutions. It is possible though to numerically estimate parameters by, for example, *Monte Carlo* methods. During Monte Carlo *sampling*, our approximation approaches the exact solution with the increase of number of samples. Of course, such stochastic approach requires significantly more calculations than analytical methods. However, setting a reasonable precision limit and employing parallelization improve the performance to the acceptable level.

Sampling can occur in several ways: by drawing uniformly with weights proportional to p.d.f., by drawing with equal weights with drawing probability according to p.d.f. or by drawing in a hybrid way, considering probability density while keeping weights rational. The rate of the approaching to the original distribution can be significantly improved by drawing samples (called *particles*) within important areas of the p.d.f. support, e.g. around probability distribution maxima. This makes latter two drawing techniques more reasonable for practical implementations.

The created particles are treated as an approximation of the original DoB distribution. Upon state estimate assessment requests, particles are propagated according to process evolution parameters and eventually converted to continuous DoB distribution for analysis convenience.

Particles tend to concentrate (or increase their weights) near to past maxima, which leads to particles or weights degeneracy. Thus, a correcting mechanism for samples is required. One of the ways is the introduction of a *resampling* step. Below we will discuss an example of particle filter algorithms with resampling – a so-called *Bootstrap Filter* variant.

### Bootstrap Filter

At a time step $k$ we have prior information $p(x_k)$ about a system attribute $X$ distributed over possible values. At first, we represent $p(x_k)$ with $N$ particles $\tilde{x}_k^i \propto p(x_k)$, where each particle is given by a weighted Dirac delta-function $w_k^i \delta\left(x_k - \tilde{x}_k^i\right)$. Hereby, we denote sampling points with $\tilde{x}_k^i$, as well as imply indices $i \in 1, \ldots, N$ and weights $w_k^i = 1/N$.

Upon receiving a new observation, we propagate our state estimate $p(x_k)$ to the observation's time moment $k+1$:

$$p(x_k) \rightarrow p(x_{k+1}|x_k) = f\left(\tilde{x}_k^i, \theta_k\right), \tag{2.32}$$

where $f$ is an evolution function (our knowledge about the system dynamics) and $\theta_k$ is a random process noise. We use the resulting *transition distribution* $p(x_{k+1}|x_k)$ for resampling (i.e. use it as the *importance function*). This variant of a more common framework *Sequential Importance Resampling* (SIR) is called bootstrap filtering.

The resampling is performed by drawing a new set of $N$ particles $\widehat{x}^i_{k+1} \propto p(x_{k+1}|x_k)$. So, the approximation transforms to:

$$p(x_{k+1}|x_k) \rightarrow \widehat{p}(x_{k+1}|x_k) = \frac{1}{N}\sum_{i=1}^{N}\delta\left(x_{k+1} - \widehat{x}^i_{k+1}\right). \tag{2.33}$$

The observation $y_{k+1}$ and sensor parameters allow us to define a likelihood function $p(y_{k+1}|x_{k+1})$. We update our resampled propagation with this function by means of *reweighting*:

$$p(x_{k+1}|y_{k+1}) = \sum_{i=1}^{N} w^i_{k+1}\delta\left(x_{k+1} - \widehat{x}^i_{k+1}\right), \tag{2.34}$$

where new weights are proportional to likelihood values at sampling points: $w^i_{k+1} \propto p(y_{k+1}|\widehat{x}^i_{k+1})$. This posterior distribution serves as the prior distribution for the next algorithm iteration.

Resampling with LDQ

There are many resampling techniques (i.e. drawing mechanisms) suggested in literature: e.g. multinomial [Smi92], residual [Hig97], systematic [Kit96], [Che03], [Afo08]. We propose an alternative to classical resampling by setting variance of resampling windows equal to corresponding least discernible quanta (refer to Sec. 3.3.3) in the parameter space. This requirement introduces physical parameters and desired task accuracy to resampling process.

### 2.3.6 Weak Distance Employment in Distribution Components Reduction and PF Sampling

*Introduction* In [Pak13] we consider distribution components reduction and particle filter sampling as follows. Employing the weak distance, introduced in Sec. 2.3.4, we can estimate the sufficient number of components required for a given problem. In the case of particle filtering, we can estimate a number of required particles and their weights. Since many PF methods rely on uniform-weight sampling (e.g. [Kot03], [Mus01]), which do not ensure optimality of the state approximation, we can expect a representation and performance improvement by employing optimal parameters.

*Definition* The Gaussian mixture reduction task corresponds to finding of another GM sufficiently good approximating the original one but with significantly fewer number of kernels. More exactly, given a distribution $f$ represented as a $M$-kernels Gaussian mixture:

$$f(P_f;\vec{\boldsymbol{x}}) \triangleq f\left(\{a_m,\vec{\boldsymbol{x}}_m,\boldsymbol{C}_m\}_{m=1}^{M};\vec{\boldsymbol{x}}\right) = \sum_{m=1}^{M} a_m \cdot G(\vec{\boldsymbol{x}}_m,\boldsymbol{C}_m;\vec{\boldsymbol{x}}), \tag{2.35}$$

where $a_m \geq 0$ are weights, $P_f$ or $G(\vec{\boldsymbol{x}}_m,\boldsymbol{C}_m;\vec{\boldsymbol{x}})$ are normalized kernels, $\vec{\boldsymbol{x}}_m$ are mean vectors and $\boldsymbol{C}_m$ – covariance matrices, we are searching for an approximating GM $g(P_g;\vec{\boldsymbol{x}}) \triangleq \sum_{q=1}^{Q} b_q \cdot G(\vec{\boldsymbol{y}}_q,\boldsymbol{D}_q;\vec{\boldsymbol{x}})$ with $Q \ll M$ and difference:

$$D_{f,g} = \arg\min_{P_g} d_T[f(P_f),g(P_g)]. \tag{2.36}$$

This minimum value of $d_T$ is a conversion loss (degradation) in terms of total integrated weight. This value gives no feeling if the number itself is big or small. We need a relative point to compare to, e.g. to the weight of the $d_T[f,z]$ with $z$ denoting a zero.

Similarly, the task of optimal sampling of original distribution $f$, is to find the best approximation with samples, e.g.:

$$h(P_h;\vec{\boldsymbol{x}}) \triangleq \sum_{s=1}^{S} c_s \cdot \delta(\vec{\boldsymbol{z}}_s - \vec{\boldsymbol{x}}). \tag{2.37}$$

During the minimization, similarly to Eq. 2.36, we are trying to find both optimal positions and weights (contrary to standard approaches). Such optimal sampling depends on the function of interest – our method is equivalent to minimization of quadratic average error in the expectation of all probe functions serving as functions of interest.

### Discrete Case

Let us continue with the example in Sec. 2.3.4. We suppose that camera measurements introduce only position-dependent Gaussian noise, i.e. the likelihood has the form: $p(\vec{\boldsymbol{\mu}}|\vec{\boldsymbol{x}}) = c_{\vec{\mu}} \cdot G(\vec{\boldsymbol{\mu}}, \boldsymbol{E}_{\vec{\mu}}; \vec{\boldsymbol{x}})$. As stated earlier, Gaussian distributions have infinite tails and, thus, seem to be poor probe function candidates. Nevertheless, it would be convenient to use closed form solutions while we are dealing with GMs and sampled distributions. Particularly, convolutions of our likelihoods and GMs, as well as derivatives of each $\langle g, \psi_l \rangle$ relative to components of $P_g$ are computable in closed form.

We minimize the value of the discretized metric in Eq. 2.31 with a gradient descent within the parameters space. Hereby, the minimization $D_{f,g}$ in Eq. 2.36 is performed with the following *weak-discrete* (WD) algorithm [Pak13]:

1. initialization
   a) choose $L$ probe function positions $\vec{\boldsymbol{\mu}}_l$ according to density $w(\vec{\boldsymbol{\mu}})$, determine weights $c_l := c_{\vec{\mu}_l}$ and covariant matrices $\boldsymbol{E}_l := \boldsymbol{E}_{\vec{\mu}_l}$;
   b) calculate $L$ convolutions $\langle f, \psi_l \rangle$;
   c) initialize $g = g^{(0)}$ (e.g. with any other reduction algorithm);
2. gradient search
   a) calculate $L$ convolutions $g_l = \langle g, \psi_l \rangle$, and their partial derivatives $\partial g_l / \partial (b_q)_i$, $\partial g_l / \partial (y_q)_i$, etc.;
   b) calculate $d_T[f,g]$ (Eq. 2.31) and its derivatives;
   c) perform gradient descent step $g = g^{(i+1)}$;
   d) stop if the approximation is appropriate for the given task, otherwise go to step 2a).

The algorithm complexity is proportional to the product of $L$, $Q$ and loop descent steps but almost invariant with respect to $M$. This is a significant difference to state of the art reduction algorithms, which usually scale as $M^2$ [Wes93].

Continuous Case

For discussion clarity, let us simplify the measurement model with the assumption that the observation accuracy is position invariant, i.e. $p(\vec{\boldsymbol{\mu}}|\vec{\boldsymbol{x}}) = G(\vec{\boldsymbol{\mu}}, \boldsymbol{E}; \vec{\boldsymbol{x}})$. Moreover, we assume that eigenvalues of $\boldsymbol{W}$ in the weight function $\Omega_\psi(\vec{\boldsymbol{\mu}}) = G(\vec{\boldsymbol{\omega}}, \boldsymbol{W}; \vec{\boldsymbol{\mu}})$ are much larger than those of $\boldsymbol{E}$. Then the weak metric is:

$$d_T[f,g] = \int \left( G(\vec{\boldsymbol{\omega}}, \boldsymbol{W}; \vec{\boldsymbol{\mu}}) \int \left(f(\vec{\boldsymbol{x}}) - g(\vec{\boldsymbol{x}})\right) G(\vec{\boldsymbol{\mu}}, \boldsymbol{E}; \vec{\boldsymbol{x}}) d\vec{\boldsymbol{x}} \right)^2 d\vec{\boldsymbol{\mu}}. \tag{2.38}$$

For clarity, we mention once again the physical meaning of the integral components: $G(\vec{\boldsymbol{\mu}}, \boldsymbol{E}; \vec{\boldsymbol{x}})$ is a Gaussian blur filter, which blurs functions of $\vec{\boldsymbol{x}}$ and produces from them smooth functions of $\vec{\boldsymbol{\mu}}$, and $G(\vec{\boldsymbol{\omega}}, \boldsymbol{W}; \vec{\boldsymbol{\mu}})$ is a windowing function of $\vec{\boldsymbol{\mu}}$, which suppresses everything outside of the window of size $\boldsymbol{W}$ (modulation process).

The algorithmic solution to the continuous case can be reduced to the ISD (Eq. 2.22) minimization problem. To this end, we start with blurring and modulating $f \to f'(P'_f; \vec{\boldsymbol{\mu}})$ by calculating internal integrals in closed form (since $f(\vec{\boldsymbol{x}})$ is representable in GM) and producing GMs in the $\vec{\boldsymbol{\mu}}$-space:

$$\begin{aligned} f'(\vec{\boldsymbol{\mu}}) &\triangleq G(\vec{\boldsymbol{\omega}}, \boldsymbol{W}; \vec{\boldsymbol{\mu}}) \int f(\vec{\boldsymbol{x}})\, G(\vec{\boldsymbol{o}}, \tilde{\boldsymbol{E}}; \vec{\boldsymbol{x}})\, d\vec{\boldsymbol{x}} \\ &= GM\left( \left\{ a'_m, \vec{\boldsymbol{x}}'_m, \vec{\boldsymbol{C}}'_m \right\}_{m=1}^{M}; \vec{\boldsymbol{\mu}} \right), \\ a'_m &= a_m \cdot G(\vec{\boldsymbol{\omega}}, \boldsymbol{W} + \boldsymbol{E} + \boldsymbol{C}_m; \vec{\boldsymbol{x}}_m), \\ C'_m &= (\boldsymbol{W}^{-1} + (\boldsymbol{C}_m + \boldsymbol{E})^{-1})^{-1}, \\ \vec{\boldsymbol{x}}'_m &= C'_m(\boldsymbol{W}^{-1}\vec{\boldsymbol{\omega}} + (\boldsymbol{C}_m + \boldsymbol{E})^{-1}\vec{\boldsymbol{x}}_m). \end{aligned} \tag{2.39}$$

The corresponding function $g'(P'_g; \vec{\boldsymbol{\mu}})$ – the result of the same transformation – is a GM with required components number $Q$ and parameters calculated by:

$$P'^{*}_g = \underset{P'_g}{\operatorname{arg\,min}} \int \left( f'(P'_f; \vec{\boldsymbol{\mu}}) - g'(P'_g; \vec{\boldsymbol{\mu}}) \right)^2 d\vec{\boldsymbol{\mu}}. \tag{2.40}$$

At this point, the task of finding of a (sub-)optimal function $g'$ is identical to the ISD minimization, which can be solved by e.g. the William's algorithm [Wil03a].

Finally, we apply the inverse transformation (un-modulate and un-blurr), converting the resulting GM into $\vec{x}$-space, as follows:

$$\begin{aligned} g(\{b_q, \vec{y}_q, \boldsymbol{D}_q\}_{q=1}^{Q}; \vec{x}) &= GM\left(\left\{b_q, \vec{y}_q, \boldsymbol{D}_q\right\}_{q=1}^{Q}; \vec{\mu}\right), \\ \boldsymbol{D}_q &= (\boldsymbol{D}_q'^{-1} - \boldsymbol{W}^{-1})^{-1} - \boldsymbol{E}, \\ \vec{y}_q &= (\boldsymbol{D}_q + \boldsymbol{E})(\boldsymbol{D}_q'^{-1}\vec{y}_q' + \boldsymbol{W}^{-1}\vec{\omega}), \\ b_q &= b_q' \cdot G(\vec{\omega}, \boldsymbol{W} + \boldsymbol{E} + \boldsymbol{D}_q; \vec{y}_q)^{-1}. \end{aligned} \tag{2.41}$$

This transformation can lead to non-positive-definite covariance matrices or nonsense weight coefficients. We mention these issues, as well as solution recipes in [Pak13].

In the following text, we will refer to the continuous algorithm version as *weak-continuous* (WC). The discussion of experimental tests and results of GM reduction is given in Sec. 4.4.3. The overall experimental results for tracking are presented in Sec. 4.4.2.

# 3

# Advanced World Modeling

## 3.1 Introduction

Many world modeling architectures and approaches discussed in previous chapters are employed in various modern intelligent autonomous systems. However, new challenges in robotics, especially in complex humanoid or multi-agent systems, demand more sophisticated frameworks. For example, modeling of arbitrary information with an open-world assumption and reasoning for decision making and cooperation in complex dynamic environments require intensive employment of sophisticated prior knowledge concepts, better classification methods, qualitative and quantitative information assessment for context analysis and so on. In this chapter we will discuss these topics in detail.

## 3.2 World Modeling Domain

### 3.2.1 Levels of Abstraction

One of possible relations (see Sec. 2.2.4) usually involved in modeling is specialization. A specialization tree, i.e. a hierarchy of concepts with the

IS A connection, allows for hierarchical classification. This hierarchy can be formed by, for example, structural criteria: specialized concepts have more attributes than a more abstract parent ([Mac10b], [Küh15]). The nodes of the specialization tree can be assigned to arbitrary defined abstraction levels. As discussed in [Ghe08] and in our works [Bel10], [Küh10] and [Bau10b] the hierarchy can be visualized as a pyramid of abstraction levels (Fig. 3.1).

Let us examine the left part of the schematics in Fig. 3.1: the Dynamic Model. At the top of the abstraction pyramid, there are blank representatives, stating existence of "something". A good example of it is presented in [Ghe08]: "if a sound is detected, it must have been emitted by a source. If no appropriate source instance can be identified in the scene model, a blank object with a suitable attribute *makes sound* is instantiated." Other examples are given in the following Sec. 3.3.3 by a discussion of interest profiles. In short, the more attribute descriptions (e.g. COLOR or SIZE) a representative has, the more concrete it becomes, lowering its level in the abstraction pyramid.

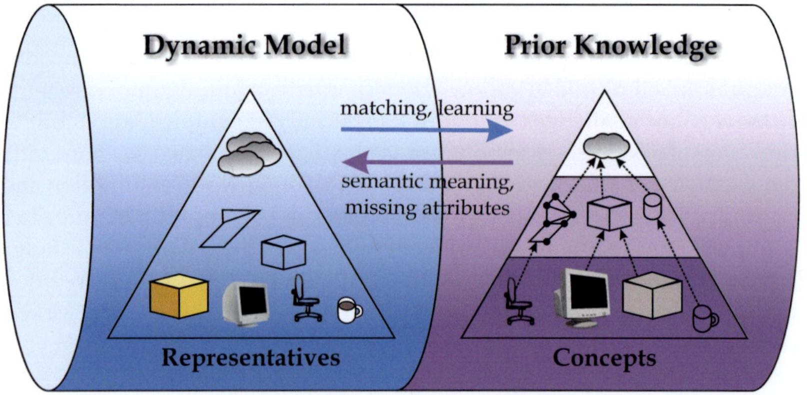


**Figure 3.1:** Schematics of a world modeling subsystem.

The right side of the schematics represents Prior Knowledge. At the top of its abstraction pyramid, there is a blank concept. It describes any possible representative. The more attribute descriptions a concept has, the lower abstraction level it takes, till the lowest level, which contains descriptions of specific classes (e.g. C++ PROGRAMMING BOOK) or even specific entities (e.g. RED COFFEE CUP OF ALEXEY).

We mention in our analyses [Bel10] and [Küh10], abstraction pyramids are naturally arising from several reasons. At first, an autonomous system knows little about the world of interest and starts to perceive entities in its proximity. Iteratively, it obtains new information pieces and creates/updates representatives, lowering their abstraction levels. The descent through hierarchy levels happens, for example, by a specialization of a representative from a more general description (e.g. CYLINDER) to a more specific description (e.g. PEN). Possible inconsistencies (e.g. PEN starts to play music) due to, for example, wrong classification can be corrected by moving a representative back to a more general non-contradictory level, e.g. PEN → CYLINDER. At the same time, each representative can be handled on each level equal or above its current, leaving unnecessary attributes out of scope [Ghe08]. For example, a table with cups on it is considered as an obstacle during a path finding (considering a corresponding bounding box only, ignoring other attributes and relations, e.g. to cups) or it is a support for cups and food during a lunch serving [Küh10]. Finally, the hierarchical abstraction pyramid is correlated with the information analysis modules pipeline. In [Küh10] we discuss such hierarchies in relation to a knowledge-driven opto-acoustic scene analysis and mention the following:

> modules at a lower abstraction level may only be executed after information from modules on higher abstraction levels is available. For example, modules for person identification cannot be executed until other modules have generated enough information, to be sure that the entity is a PERSON with a DoB above a certain threshold.

Such knowledge-driven modular scene analysis works iteratively with abstraction levels in a natural way.

## 3.2.2 Information Interchange

Since the world modeling domain consists of dynamic model and prior knowledge, it is important to ensure information interchange between these two components. This implies matching of dynamic information containers to pre-defined concepts, fetching semantic meaning and missing attributes, extending persistent knowledge by learning processes. In the following sections we discuss each of these two components, as well as information interchange.

## 3.3 Dynamic Model

### 3.3.1 Conventional Object-Oriented Approach

Dynamic representatives are usually instantiated from prior knowledge concepts (e.g. [Ghe08]). In practice, during object-oriented analysis and design we define a set of relevant classes [Boo93] (correspond to our concepts) and instantiate by them objects (correspond to our representatives) as soon as objects were classified. However, this approach is affected by several problems (refer to our work [Bel10]):

1. multiple hierarchies – we can build many specialization hierarchies based on different distinctions, e.g. on geometrical form (e.g. BLANK CONCEPT → CYLINDER → CUP) or on functional assignment (e.g. BLANK CONCEPT → KITCHEN ENTITY → CUP). Hereby, it is not clear which hierarchy to choose or how to organize multiple inheritance and typisation;

2. succession order – the system obtains different sequences of information pieces, like in one case “shape” before “temperature” (visible distant entity), in another case “temperature” before “shape” (huge complex entity nearby). It is not clear then which attribute should precede which on the levels of abstraction;

3. fixed construct limitations – a fixed set is more rigid and limited compared to symbolic descriptions [Ahl02]. This means, it is possible to classify entities according to pre-defined concepts and on the basis of expected attributes only. Unexpected cases (e.g. unknown entities) cannot be flexibly handled;

4. information transfer – constant transfer of existing information of more general representative to more specific representative upon

instantiation raises additional technical complications (e.g. memory management, additional processing operations within an aspired real-time framework, etc.);

5. physical limitations – a set of concepts has to be limited due to development and maintenance limitations, as well as due to memory and CPU-time consumption. Additionally, large sets increase mismatch rate due to a large number of concepts with similar structure. On the other hand, real-life environment contains an arbitrary large number of arbitrary concepts (open-world assumption).

The latter issue can be solved by introduction of thematic modules. Each module contains information relevant to some scene (e.g. to kitchen or living room) and context (e.g. to party or siesta). The world modeling subsystem loads and unloads such modules on demand (e.g. by situation recognition subsystem). Due to the first four issues, the conventional object-oriented approach represents a technical workaround limited to a narrow scope of tasks. We propose a better method in the next section.

### 3.3.2 Progressive Mapping

The *Progressive Mapping* mechanism was introduced in our works [Bel10] and [Küh10] as follows: each blank representative is created as an empty information container. Upon first measurement of an attribute, the modeling subsystem creates a corresponding attribute description and maps it to the representative. This mechanism eliminates first four issues listed in the Section 3.3.1. Namely, instead of fixed constructs we operate with dynamic flexible structures. The incoming information is fused into existing mapped attribute description with consistency checks. A temporal progression of a representative's description is presented in Fig. 1.3.

### 3.3.3 Quantitative Assessment

In [Bel12b] we propose a quantitative assessment for numerical estimation of uncertainty within situations at hand. This mechanism can be extended for a general uncertainty assessment of the dynamic model, i.e. how well are attributes, entities and situations known. Moreover, it is possible to assess the sufficiency of the information relative to a given task. To this end, we introduce Shannon (information) entropy to our calculus.

### Information Entropy

A common way to determine the average amount of information contained in world models or incoming observations is to calculate the information entropy [Cov91]. Namely, the entropy describes uncertainty about our source of information. We introduce the entropy with the following formulas:

$$H(A) = E[-\log_2 p(a)] \qquad (3.1)$$
$$\Rightarrow \begin{cases} H(A) = -\sum_{a \in D_A} P(a) \log_2 P(a), \\ h(A) = -\int_{a \in D_A} p(a) \log_2 p(a) da, \end{cases}$$

where $A$ denotes an attribute and $a$ its discrete or continuous value from a $D_A$ definition set.

Incoming observations of the surrounding world reduce uncertainty quantity in our dynamic model of the surrounding environment, thus, reducing the value of corresponding entropy. The resulting change in the information quality is the difference of an entropy of a previous estimate and an entropy of an updated estimate:

$$M_k = H_{k|k} - H_{k|k-1}. \qquad (3.2)$$

### Entropy Unification with LDQ

We have both discrete and continuous probability distributions within our modeling subsystem, which we treat equally. Thus, the entropy calculation of attribute descriptions implies equal treatment of discrete and continuous pdfs. However, the entropy of a p.d.f. differs from the entropy of a p.m.f. in nature. In [Bel12d] we point out that even discretization of an attribute $A$ to $A^*$ with smaller and smaller steps ($P(a^*) \to p(a)$), does not solve this problem. In this case, we obtain a divergence of the discretized entropy to the continuous one, so that $\lim_{a^* \to a} H(A^*) \neq h(A)$. Namely, the entropy of an $n$-segments p.d.f.-discretization is approximately $h(A) + n$, which leads to infinite discrepancy in the case of $n \to \infty$. We overcome this issue by introducing the notion of a least discernible quantum (LDQ) [Osw91], which defines the maximum precision of all operations over the given attribute. The quantum size $\Delta$ (i.e. maximum precision) is determined by e.g. sensor acceptance and efficiency (physical limitations), as well as task requirements (required information granularity).

We propose the following entropy unification mechanism: the entropy for discrete distributions is calculated according to the standard Equation 3.1, while continuous distributions are pre-discretized by integrating of $\Delta$-slices (i.e. histogramming procedure) [Bel12b], as presented in Fig. 3.2:

$$\begin{aligned}
&a \mapsto \text{slices } v,\\
&v \in D_A^\Delta \subseteq \mathbb{N},\\
&a \in \left[a_v - \frac{\Delta_A}{2}; a_v + \frac{\Delta_A}{2}\right],\\
P(a) \to P(v) &:= \int\limits_{a_v - \frac{\Delta_A}{2}}^{a_v + \frac{\Delta_A}{2}} p(a)da,\\
H_\Delta(A) := H(v) &= - \sum_{v \in D_A^\Delta} P(v) \log_2 P(v).
\end{aligned} \tag{3.3}$$

Hereby, we distinguish two extreme cases for the discretized entropy: the attribute's uncertainty is less than LDQ and the attribute's value is completely unknown. In the first case, $a$ is known better than $\Delta_A$, which means that the complete support of the probability distribution is within $\Delta_A$, as presented in Fig. 3.3a. Then we have:

$$H_\Delta(A) = 0. \tag{3.4}$$

If $A$ is completely unknown, then $P(v) = const = c_v$ as presented in Fig. 3.3b and $c_v \cdot |D_A^\Delta| = 1$ (normalized p.d.f.), then:

$$H_\Delta(A) = - \sum_{v \in D_A^\Delta} c_v \log_2 c_v = - \underbrace{c_v \cdot |D_A^\Delta|}_{1} \log_2 \underbrace{c_v}_{1/D_A^\Delta} = \log_2 |D_A^\Delta| = H_\Delta(A)_{\max}. \tag{3.5}$$

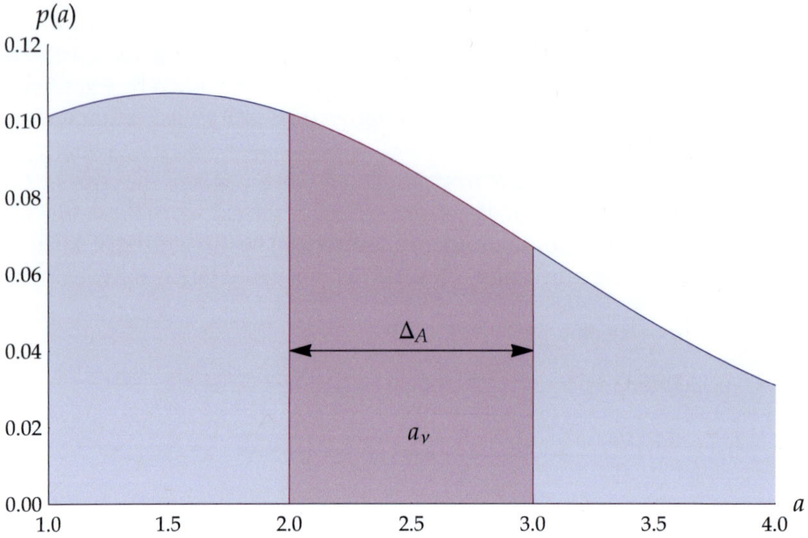


**Figure 3.2:** Discretization of a p.d.f. by LDQ $\Delta_A$.

For correct estimation, the maximum entropy value is calculated from the attribute description's value definition range and not from an arbitrary distribution.

Since we have now a unified mechanism for entropy calculation of all attribute descriptions, we can estimate the uncertainty of modeling information. From practical considerations, we introduce a new notion for this.

### Bestimmtheit

The entropy calculation allows for estimation of uncertainty for attribute descriptions. For convenience, we define the opposite value, a so-called

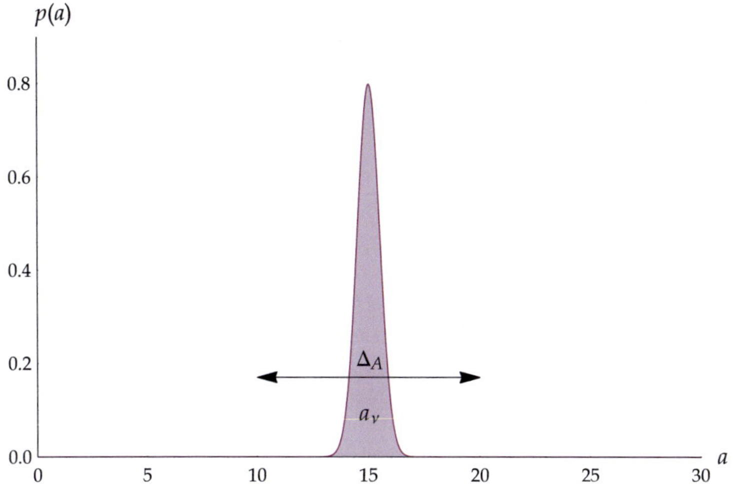


**(a)** Peak within the LDQ quantum;

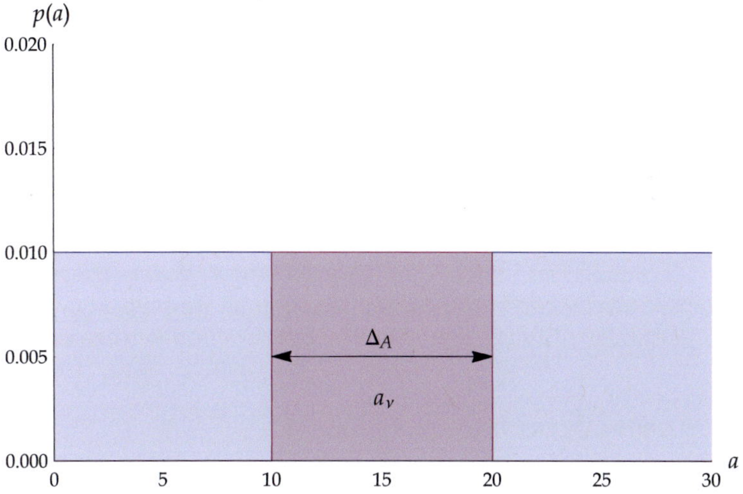


**(b)** Flat p.d.f.;

**Figure 3.3:** Discretization of a p.d.f. by LDQ $\Delta_A$: extreme cases.

*Bestimmtheit* (German word for "determinedness") $B$, characterizing the degree of sufficiency in the given information [Bel12b], as follows:

$$B : P(a) \to [0,1],$$

$$B(A) := 1 - \frac{H(A)}{H(A)_{\text{max}}}, \tag{3.6}$$

$$B_\Delta(A) := 1 - \frac{H_\Delta(A)}{H_\Delta(A)_{\text{max}}} = 1 - \frac{H_\Delta(A)}{\log_2 D_A^\Delta}. \tag{3.7}$$

Naturally, the extreme cases for Bestimmtheit are: $A$ is sufficiently known ($B_\Delta(A) = 1$) and $A$ is completely unknown ($B_\Delta(A) = 0$).

### Bestimmtheit Assessment of Dynamic Models

Since we can calculate Bestimmtheit for any attribute description in our dynamic model, it is reasonable to assess representatives or situations as a whole, in order to estimate modeling sufficiency for entities of interest. An even more useful measure could be the sufficiency relative to given tasks or goals. To this end, we have introduced in [Bel12b] a weight function that expresses the relative importance of attributes (e.g. assigning more weights to spatial attributes by path finding):

$$w(A) : A \to [0,1], \tag{3.8}$$

$$\sum_{A \in \{A\}_r} w(A) = 1,$$

where $\{A\}_r$ denotes attributes of the representative $r$. If no task is provided, we suppose this function to be e.g. uniform over all attribute descriptions.

The assessment of a given representative can be performed then as follows:

$$B(r,w) := \sum_{A \in A_r} w(A) B_{\Delta_A}(A). \tag{3.9}$$

Similarly, it is possible to assess situations (or, in principle, any element of the dynamic model):

$$B(\{r\}, w_r, \{w\}) := \sum_{r\in\{r\}; w\in\{w\}} w_r(r) B(r, w), \tag{3.10}$$

where $\{r\}$ and $\{w\}$ denote representatives and weight functions respectively.

In [Bel12b] we mention also an *overdominance* problem for the Bestimmtheit calculus. For example, if some task requires 5 attributes to be known but only 4 of them are observed, then this information is not adequate for the task accomplishment. However, Bestimmtheit values of these 4 attribute descriptions can numerically "compensate" the missing one (so, overdominate), leading to "sufficient" $B$. The overdominance problem can be solved by introduction of nonlinear weighting function, as well as a threshold for each attribute: $B_\Delta(A) \geq B_{\Delta,\min}(A)$.

### Interests Profile

The Bestimmtheit assessment allows for numerical estimation of degree of sufficiency of dynamic model contents. As we mention in [Bel12b], such assessment reveals the information deficit relative to given tasks (i.e. to crucial attributes and representatives, which in turn form so-called "interests profile"). This can be used for triggering of actuators (e.g. for sensor control planning, active vision, etc.) or cognition modules (interactive learning, questioning in human-machine interactions, information acquisition optimization). Formally, we can define a function $f(H, H_{\max}, \Delta)$ that assesses the information sufficiency relative to required precision.

Additionally, the interests profile can be extended by spanning blank representatives onto unexplored area volumes. This assures that the autonomous system will not count these volumes as free [Ghe08]. Hereby, an autonomous system can switch into an "exploration" mode, adjusting "interest" weights to blank representatives according to the distance to them. This will stimulate the system to explore the surroundings, starting from the proximity. Similarly, we can adjust weights of interesting attributes and representatives by a salience-based exploration ([Küh15]).

### 3.3.4 Qualitative Assessment

We have already discussed the qualitative assessment of the dynamic model contents (e.g. on the example of situations assessment) in Sec. 1.2.3 and in [Bel12d]. Here, we imply a symbolic interpretation of the dynamic model in order to infer context descriptions (for a detailed introduction into this subject refer to e.g. [Bau13]). This includes also a temporal assessment for finding present situations at given scenes (for situations matching refer to our work [Bel12b]). As we mention in [Bel12b]:

> Each situation at hand can be compared to pre-defined set of triggers by given representatives, attribute and relation descriptions. For example, if there is a group of people drinking alcohol and singing “Happy Birthday”, then it can be a birthday party. The prediction can be handled with a Hidden Markov Model. The assessment of the situation at hand can be evaluated as as described in Section 3.5.2 – within a common framework for the prior knowledge matching.

Such qualitative assessment affects the whole process of decision making and planning. For example, a robot postpones vacuum cleaning if a situation PERSON SLEEPING is found.

To enable situation recognition and decision making processes it is also convenient to form a Bayesian Network from the dynamic model contents, which is a separate vast research area ([Fis15]).

### 3.3.5 Relevance

Relevance mechanisms are responsible for delivering only relevant information to querying sub-modules, as we have mentioned it in Sec. 3.2.1. This includes selection of the appropriate abstraction level and sometimes preprocessing of information. For example, cognition sub-modules can query information about all entities on the kitchen table, including type and state attributes, in order to assess the situation at present (e.g. find out if there is a breakfast scenario). In contrary, during the path planning, cognition modules require only geometrical information of entities, namely bounding boxes.

### 3.3.6 Consistency

Consistency mechanisms are responsible for finding and correcting model errors. For example, wrong classification can be corrected by moving representative to a more general level: e.g. if a PEN starts to play music, it can be reclassified by moving the representative to a more general level, e.g. PEN → CYLINDER, as mentioned in Sec. 3.2.1. Another example – multiple representatives created and matched to the same entity due to sensor uncertainty. This can be corrected by representatives fusion triggered by temporal analysis (several representatives to the same entity over some time duration) of the scene.

## 3.4 Prior Knowledge

### 3.4.1 Intrinsic Prior Knowledge

As we have already mentioned in Chapters 1 and 2, there is intrinsic prior knowledge, such as methods for sensory information processing, information association and fusion, evolution models and inference processing. Additionally, intrinsic prior knowledge contains directly or indirectly a global scheme of attributes, describing known types of attributes. Such description contains meta-information for each attribute: allowed values, precision, coordinate system with reference, color space, etc. In principle, the intrinsic prior knowledge about attributes brings semantics to attribute descriptions. In the same way, prior knowledge about entities brings semantics (e.g. TYPE attribute) to representatives, which is required for cognition and planning modules.

### 3.4.2 Topology

The environment topology information is a part of prior knowledge. It can be pre-defined and/or dynamically created during the exploration process. Pre-defined information can be created by 3D laser scanner system (e.g. as depicted in Fig. 4.7). A dynamic creation of topology maps is performed usually during SLAM.

### 3.4.3 Ontology

If the world of interest is limited (for example, to a laboratory environment), the choice of representation of concepts is usually not crucial. Complex environments (in extreme case an open-world one) require more general representations. [Rus10] suggests for this the ontology representation:

> we won't actually write a complete description of everything ... we will leave placeholders where new knowledge for any domain can fit in. For example, we will define what it means to be a physical object, and the details of different types of objects – robots, televisions, books, or whatever – can be filled in later ... The general framework *upper ontology* of concepts is called an upper ontology because of the convention of drawing graphs with the general concepts at the top and the more specific concepts below them ...

In contrast to ontology in philosophical study means, we understand under the information science notion "ontology" a formal explicit description of a domain under study in terms of concepts and their connections. This implies naming and definition of types, attributes and relations of entities of the world of interest. A good introduction into ontology-based unified knowledge representation for robots is given in [Lim11].

We assume that prior knowledge ontology contains:

- concepts of known classes (e.g. CUPS, TABLES);
- concepts of specific entities (e.g. RED COFFEE CUP OF ALEXEY);
- rules and restrictions;

As we mention in [Bel12a], each concept contains a set of attribute and relation descriptions in form of prior probability distributions, as well as semantic specifications. These distributions can contain also real statistical information as discussed in [Kas13]. For example, it can contain HEIGHT pdf for the concept COFFEE CUP based on a statistical analysis over all coffee cups in a household. In the absence of statistical information, prior knowledge distributions can be flat over the allowed attribute's definition range. However, this leads to unwanted effects, e.g. since distribution tails of representative's attribute descriptions usually exceed prior knowledge distributions' boundaries, representative-to-concept matching with Kullback-Leibler divergence delivers infinite difference at tail areas.[1] Thus, we prefer to represent concept's attributes in form of Gaussian distributions or mixtures. Hereby, equal representation of attribute descriptions of representatives and concepts leads to universal representation and processing framework and facilitates development of the system. Moreover, Gaussian form allows pre-setting of most probable value(s). Even without statistical information, ontology experts can choose such values by reasoning about personal experience, e.g. setting 4 cm value as mean and more or less matching variance for radius distribution of tea cups (meaningless tails, for example, values below zero, are treated according to the intrinsic knowledge convention).

The semantics, i.e. symbolic descriptions, is important for cognitive processes. It can be introduced in different ways [Lim11], e.g. by predefining by experts. The semantic specifications can include information, such as CAN CONTAIN LIQUIDS, HAS A STABLE SURFACE ON THE BOTTOM and BELONGS TO KITCHEN for further qualitative assessments.

An ontology structure depends on the complexity of the domain to be modeled. However, even a simple environment leads to a sophisticated graph of concepts and their relations of different types (Figure 3.4). Some relations, e.g. IS A, form trees of semantically consistent hierarchies. These hierarchy trees are coherent to abstraction levels defined in Section 3.2.1. Moreover, we can make the structure more formal and always provide the normalized total probability, which represents probability distribution over all possible concepts, by supplementing each branch with a "dummy" concept [Bel12a]. These dummy concepts include all other types, except those that are already listed, e.g. CUP → {COFFEE CUP, TEE CUP, DUMMY}.

In the Figure 3.4 we can see two IS A hierarchies: functional (solid blue and violet lines) and geometrical (solid green lines). Moreover, concepts are connected by dependency links (dashed violet lines). We can group concepts into modules, e.g. all things related to kitchen are organized into

1 The KL divergence is asymmetrical. Thus, an appropriate prior knowledge distribution definition combined with an appropriate choice between the KL divergence of the $(p,q)$ and the $(q,p)$ can help to avoid this problem.

a KITCHEN MODULE. This is the practical realization of thematic modules mentioned in the Section 3.3.1.

Additionally, we can consider tree subbranches as partitioning of the parent branch (more abstract concept) into particular concepts (e.g. FURNITURE $\rightarrow$ {TABLE, CHAIR}) as depicted in Fig. 3.7b.

### 3.4.4 Complex Entities

It is important to model complex entities, e.g. those consisting of multiple parts (e.g. BOTTLE OF WINE: {CORK, GLASS BOTTLE, WINE} and entities of complex geometry. Modeling of such entities is covered in a parallel project within the DFG SFB 588 [Kas13].

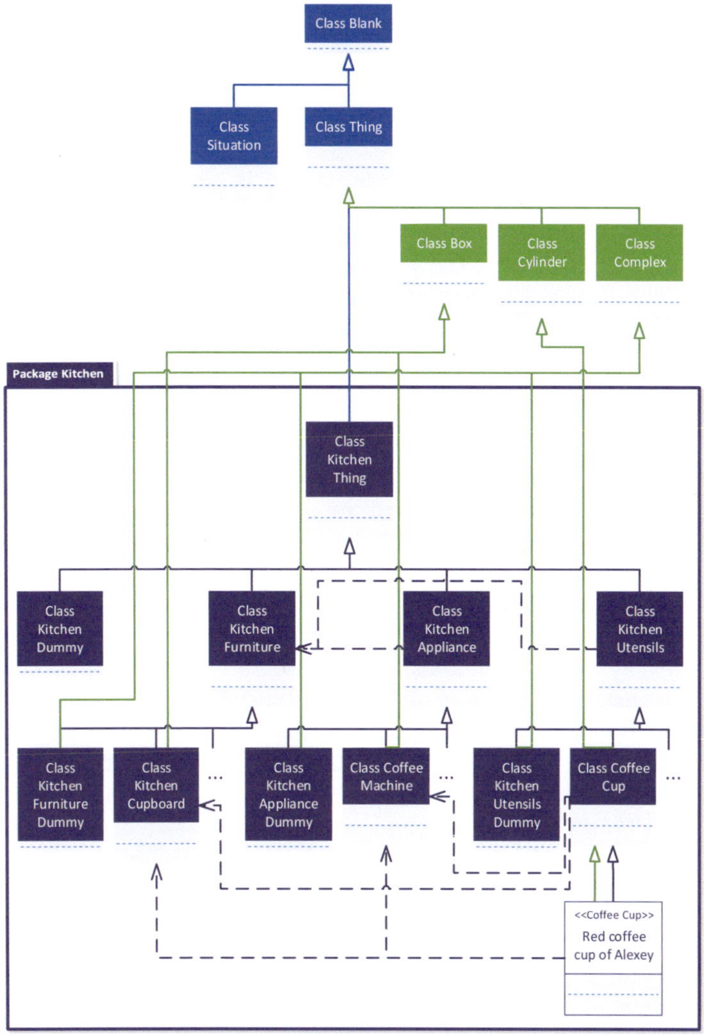


**Figure 3.4:** Ontology of two IS A hierarchies: solid blue and violet lines are functional and solid green lines are geometrical connections. Dependencies between concepts are depicted by dashed lines.

## 3.5 Connection of Dynamic Models to Prior Knowledge

Before being connected to prior knowledge, all representatives are abstract information containers with mapped attribute descriptions. In order to bring semantic meaning to them, we need first to classify representatives, i.e. match these dynamic containers to known concepts. A good introduction into classificator subjects is given in e.g. [Mac10b]. As stated in our work [Bel10], if an entity is recognized, for example, as a cup, a corresponding representative can get information from prior knowledge ontology that the entity e.g. can be used to carry liquid, has a stable surface on the bottom and belongs to kitchen. In [Bel12c], we discuss a "table example":

> if the assigned concept is TABLE, the autonomous system is informed that it is used as a placeholder for other entities – as an IS ON surface with corresponding relations to other representatives. This information can be used further by assigning well-known TABLE HEIGHT to all entities laying on it. Second, the robot is informed that this surface can be used for placing things onto it. Third, this surface can be considered further as a search location. The knowledge about this location can be extended by a statistical distribution of how often and where exactly specific things are found there. Later on, it can be used for a heuristic search, e.g. by the command BRING ME THE APARTMENT KEYS AND MY MOBILE PHONE.

Moreover, an extended set of attributes (e.g. probability mass function MADE OF) pre-defined in prior knowledge can be transferred to corresponding representatives upon successful matching. A general discussion of information interchange between dynamic models and prior knowledge is presented in our works [Bel11], [Ghe10], [Bel12d] and [Bel12c].

### 3.5.1 Inference

In [Bel10] and [Küh10] we discuss a possibility to infer missing attributes. Let us suppose that an autonomous system finds a ball of approximately 23-24 cm diameter and of a distinct black-white pattern (truncated icosahedron). The modeling system matches this entity to known concepts, classifies it and maps a new attribute description TYPE to the ball representative. In this example, the type p.m.f. $P(c|r)$, where $c$ denotes class and $r$ representative, has non-zero values at FOOTBALL $P(c=F|r)$, VOLLEYBALL $P(c=V|r)$ and other types. Now we can deduce the attribute WEIGHT as follows:

$$P(w|r) = \sum_X P(w|c=X,r)P(c=X|r). \tag{3.11}$$

Similarly, we can infer information by means of whole representatives, situations and contexts.

### 3.5.2 Matching to Prior Knowledge

#### Introduction

In order to employ an ontology of concepts within our dynamic modeling, we have to dynamically link prior knowledge database elements to representatives. To this end, we can try to match representatives and scenes against ontology contents, categorize them and extend with semantic meaning and missing attributes. In [Bel12a], we propose a structural/value matching of representatives to concepts. This matching is discussed in [Bel12c] in details and demonstrated by an experiment of exploration of entities on a table during a coffee break. Later, we extend dynamic model to prior knowledge matching with situations estimation in [Bel12b]. The discussed ideas are as follows: since matching of a representative to a complex ontology graph is computationally hard, we consider only one generalization hierarchy at time. For example, let us consider the ontology in Figure 3.4 and take the geometrical IS A hierarchy. Then, the remaining graph would as in Figure 3.5. Next, we visit tree nodes (e.g. with a *depth search*), finding concepts similar to the given representative. The lowest concepts that get similarity score higher than a threshold we mark as appropriate candidates, creating a set $\{C\}_r$. At the end of the matching, we have marked branches as e.g. in Fig. 3.6 that give us type probability distribution presented in Fig. 3.7a. This kind of generalization hierarchies is, in principle, a partitioning of type probabilities (Fig. 3.7b).

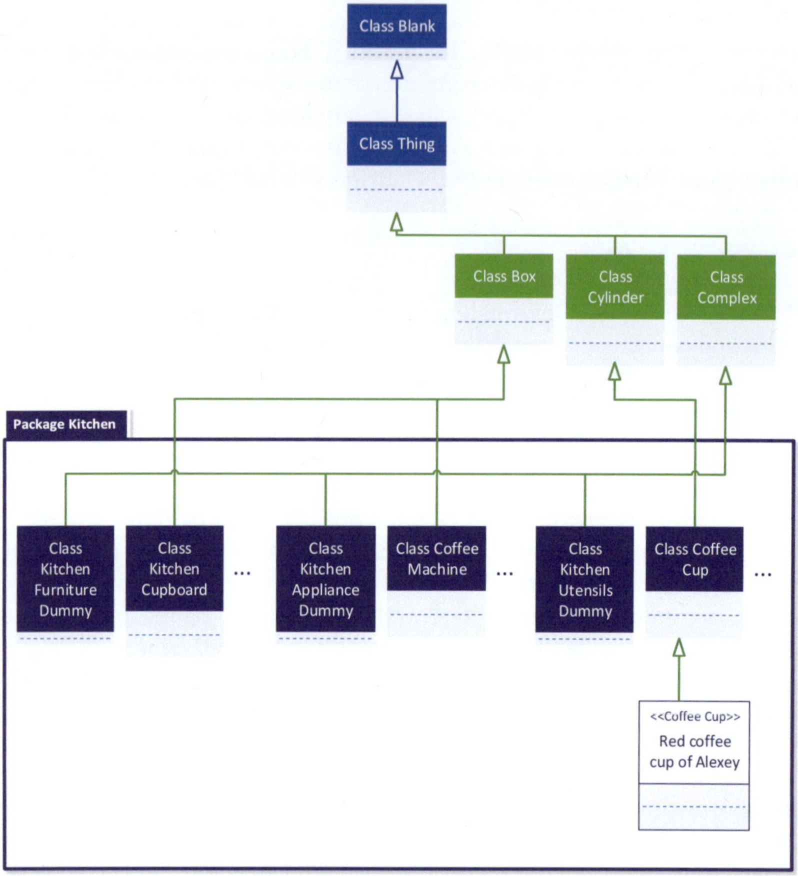


**Figure 3.5:** Geometrical IS A hierarchy.

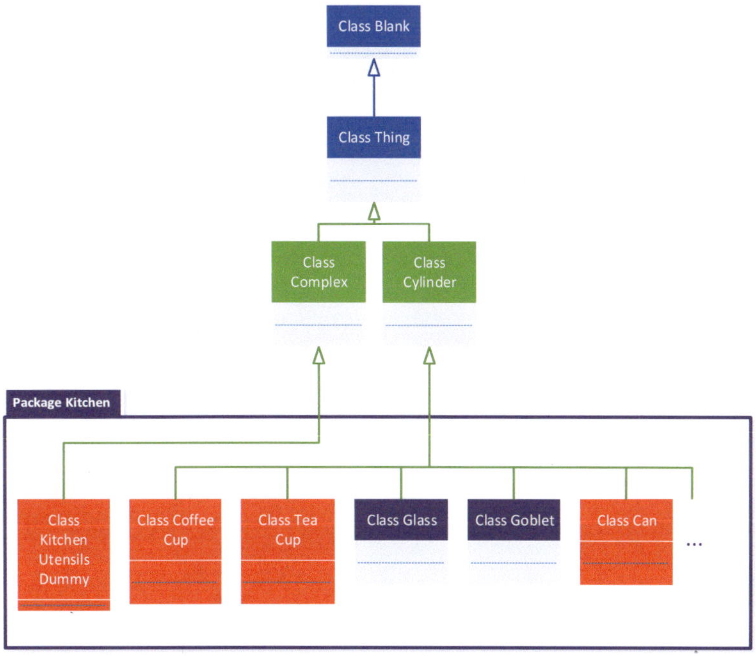


**Figure 3.6:** Matching result: found candidates (red).

During the similarity computation, we have to consider structural and value similarities of prior knowledge concepts to the given representative [Bel12a]. A normalized attributes intersection, i.e. how many known attributes of a concept were measured for the given representative, gives the structural similarity. From the other side, pdfs comparison gives value similarity over attribute descriptions at hand. The required type probability (for example, as in Fig. 3.7a) we denote as $P(c|r)$ (the probability of concept $c$ given representative $r$, i.e. given its attribute descriptions set), where $r$ is the representative under consideration and $c \in \{C\}_r$ are concepts marked as appropriate candidates. Within the Bayesian formalism, we can express this

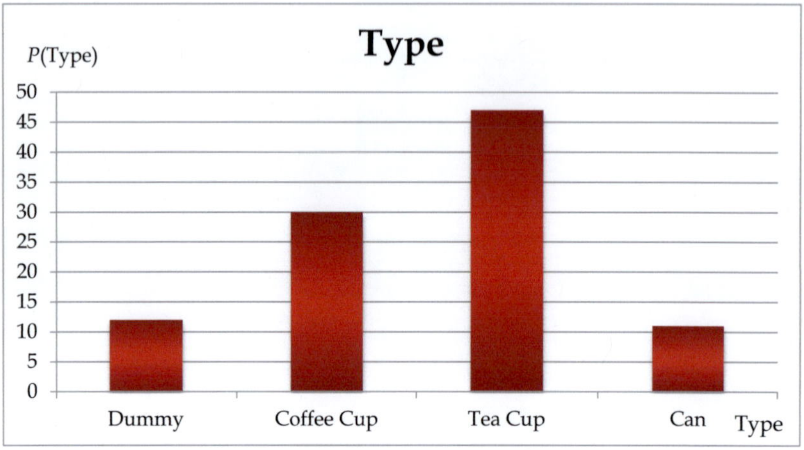


**(a)** Resulting TYPE distribution;

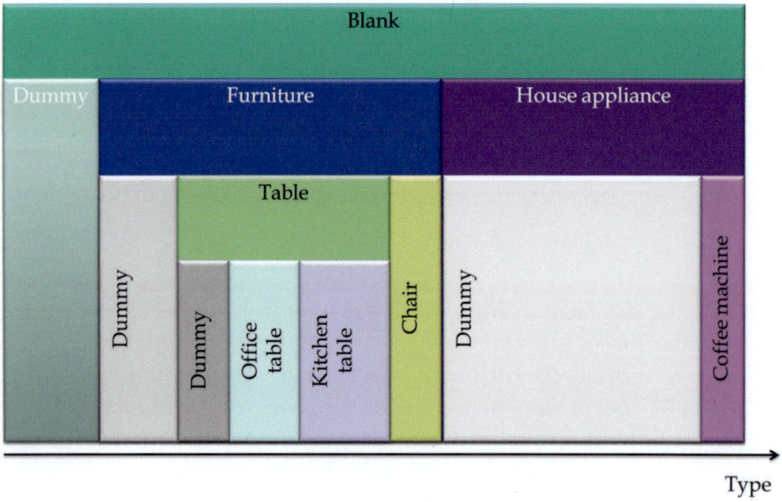


**(b)** Partitioning of the normalized total probability;

**Figure 3.7:** Matching to prior knowledge.

probability as follows:

$$P(c|r) = \frac{P(r|c)P(c)}{P(r)} \tag{3.12}$$

The $P(c)$ is the prior knowledge, i.e. how often we meet class $c$, and $P(r)$ is the probability that we meet representative $r$. The likelihood $P(r|c)$ is the point where we perform the structural and value comparison with some metric $d(r,c)$ that quantifies the difference between $r$ and $c$. We also normalize structural and value distances with a functional $f$ that projects distances onto the range $[0;1]$. The formalism is then:

$$P(r|c) := d(r,c), \tag{3.13}$$

$$d(r,c) = \lambda_s \underbrace{f_s\,(d_s(r,c))}_{\text{structural difference}} + \lambda_v \underbrace{f_v\,(d_v(r,c))}_{\text{value difference}}, \tag{3.14}$$

$$f_x\,(d_x(r,c)) : [0;\infty] \rightarrow [0;1],\; x \in \{s,v\} \tag{3.15}$$

$$\sum_{c \in C_r} f_x(d_x(r,c)) = 1,$$

where $\lambda_s$ and $\lambda_v$ are weighting factors. The choice of the functional $f$ is a matter of design (e.g. $f(d) = 1 - e^{-d}$) and eventually task relevant.

There are many different metrics both for structural and value comparison, e.g. described in [Cha07], [Wol81], and [Gib02]. In the following sections, we will discuss the most popular of them.

### Structural Difference

The structural similarity implies counting attributes that exist in a dynamic description, finding same attributes in ontology concepts and calculating the overlap value (Fig. 3.8). Practically, such comparison can be performed by metrics for sets, e.g. Tanimoto distance [Rog60], Sørensen index [Sør57] or other metrics discussed in the overview articles [Wol81] and [Jac89]. In total, there are more than 30 structural distances discussed in the literature, mainly coming from the fields of biology or geology. In this section we discuss three popular distances: Tanimoto, Jaccard and Sørenson.

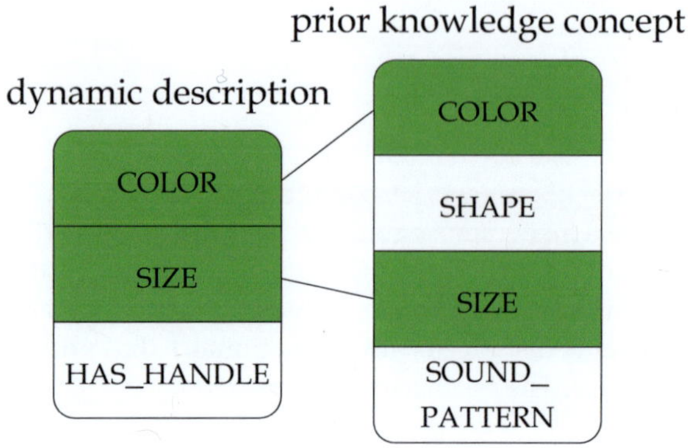


**Figure 3.8:** The structural similarity implies counting attributes that exist in a dynamic description, finding same attributes in ontology concepts and calculating the overlap value.

As we mention in [Bel12c], the Tanimoto distance is widely used for comparison of finite sets (e.g. structural similarity search for bio-molecule classification [Kar06] or land cover detection [Yan10]) but the connection of prior knowledge to dynamic modeling of autonomous systems has been not analyzed up to now. The same applies to popular Jaccard and Sørenson measures. These three distances are defined as follows:

$$d_{\text{s:Tanimoto}}(r,c) = \frac{|\{A\}_r| + |\{A\}_c| - 2|\{A\}_r \cap \{A\}_c|}{|\{A\}_r| + |\{A\}_c| - |\{A\}_r \cap \{A\}_c|}, \quad (3.16)$$

$$d_{\text{s:Jaccard}}(r,c) = \frac{|\{A\}_r \cup \{A\}_c| - |\{A\}_r \cap \{A\}_c|}{|\{A\}_r \cup \{A\}_c|}, \quad (3.17)$$

$$d_{\text{s:Sørenson}}(r,c) = \frac{2|\{A\}_r \cap \{A\}_c|}{|\{A\}_r| + |\{A\}_c|}, \quad (3.18)$$

where $|\{A\}_r|$ and $|\{A\}_c|$ are numbers of attribute descriptions in $r$ and $c$ respectively. $|\{A\}_r \cap \{A\}_c|$ and $|\{A\}_r \cup \{A\}_c|$ are numbers of overlapping and combined attribute descriptions respectively. [Wol81] states that Tanimoto distance, Jaccard and Sørensen indices are effectively equivalent.

It is important to note, though, that different distances are calculated in different scales. For example, Tanimoto-metric gives a difference of two structures (0 if same structures and 1 if totally different), while Sørensen index delivers similarity of two structures (1 if same structures and 0 if totally different). From now on, we use the difference of two structures, employing either Tanimoto-metric or inverse Sørensen index:

$$d_{\text{s:1-Sørenson}}(r,c) = 1 - d_{\text{s:Sørenson}}(r,c). \tag{3.19}$$

Value Difference

The value distance $d_v(r,c)$ is computed over attribute descriptions common to both:

$$d_v(r,c) = \sum_{A \in \{A\}_r \cap \{A\}_c} f_{DoB}(d_{DoB}(A)), \tag{3.20}$$

where $f_{DoB}$ is a normalizing functional for the attribute $A$ and $d_{DoB}(A)$ is a distance between $p(A_r)$ and $p(A_c)$ of the attribute $A$ with values $a$.

For the distance $d_{DoB}(A)$ calculation, we can find over 30 different metrics discussed in the literature, e.g. Hellinger distance, Bhattacharyya distance or Lévy-Prokhorov-Metric. One of the most popular is the Kullback-Leibler (KL) divergence [Kul51], [Cov91]. The KL divergence of $Q$ from $P$, denoted $d_{KL}(P,Q)$, is a measure of the information loss (i.e. coding penalty) when $Q$ is used to approximate $P$ [Cov91], [Bur02]. The KL divergence is also called *relative entropy,* since it is a generalization of Shannon entropy – a relative information entropy with the consideration of a model $Q$, i.e. direct employment of a prior knowledge. In other words, we search the model $Q$ that minimizes the information loss – the minimum distance with $Q$-variation by fixed (given) $P$. The KL divergence effectively measures the average likelihood of observation data $P$ if a particular model $Q$ has generated this data. It is a measure of "discrepancy", a so-called "directed measure", and not a simple distance, since KL from $P$ to $Q$ is not the same as KL from $Q$ to $P$. The KL divergence is one of the most fundamental of all information measures since it is derived from minimal assumptions and fulfills the additivity property. As we state in [Bel12c], the Kullback-Leibler distance is widely used in classification (e.g. heart signals classification [Chu08], similarity of ontology elements [Zam11] or word clustering in text classification [Dhi02]).

The KL divergence is calculated as follows:

$$d_{DoB:KL}(P,Q) = \sum_a P(a) \log \frac{P(a)}{Q(a)}, \tag{3.21}$$

$$d_{DoB:KL}(p,q) = \int_{-\infty}^{\infty} p(a) \log \frac{p(a)}{q(a)} da, \tag{3.22}$$

where $P$ and $p$ denotes posterior and $Q$ and $q$ prior degree-of-belief distributions for discrete and continuous cases respectively. Further on, we use a convention [Cov91] as follows:

- $0 \log \frac{0}{0} = 0$;
- $0 \log \frac{0}{q} = 0$;
- $p \log \frac{p}{0} = \infty$.

Among several drawbacks of KL divergence (e.g. distributions must be not Dirac functions sampled), for the GM case it is neither analytically tractable, nor any efficient computational algorithm exists [Her07], [Jen07]. In this case, one can replace GMs with suitable substitutions, like Gaussian, matched bound or variational approximations, as discussed in [Her07]. Hereby, we can employ Monte Carlo methods for achieving an arbitrary precision for the distance assessment.

Alternatively, the value difference can be calculated with a normalized Wasserstein metric $L^2$ [Haz01], [Jen07] (a variation of integrated squared difference is already discussed in 2.3.4), which has a closed form expression for any number of GM components ([Ahr05]):

$$d_{DoB}(A) = d_{L^2}(p_1, p_2) = \int \left(\acute{p}_1(a) - \acute{p}_2(a)\right)^2 da, \tag{3.23}$$

$$\acute{p}_i = \frac{p_i(a)}{\sqrt{\int p_i(a)^2 da}}.$$

A comparison of Kullback-Leibler and normalized L2 distances (along with earth movers distance) is discussed in [Jen07] applied to computational music for instruments recognition. The comparison outcome is as follows [Jen07]:

> all three distance measures perform approximately equal when using a single Gaussian with full covariance matrix, except that the normalized L2 distance performs a little worse when mixing instruments from different sound fonts. Using a mixture of ten diagonal Gaussians generally decrease recognition rates slightly... For ten mixtures, the recognition rate for the Kullback-Leibler distance seems to decrease less than for the EMD and the normalized L2 distance. From these results we conclude that the cosine distance performs slightly worse than the Kullback-Leibler distance in terms of accuracy. However, with a single Gaussian having full covariance matrix this difference is negligible, and since the cosine distance obeys the triangle inequality, it might be preferable in applications with large data-sets.

Hereby, the advantage of triangle inequality ($d(p_1, p_2)+d(p_2, p_3) \geq d(p_1, p_3)$) employment is concealed in distances computation. For example, by nearest neighbor search in "each to each" case, finding minimum distance to $p_2$ with known distances $d(p_1, p_2)$ and $d(p_1, p_3)$ is optimized as follows:

1. the distance to candidate $p_3$ is bounded by $d(p_2, p_3) \geq d(p_1, p_3) - d(p_1, p_2)$;
2. if the current minimal distance $d(p_n, p_2)$ less than $d(p_1, p_3)-d(p_1, p_2)$, we can reject $p_2$ without $d(p_2, p_3)$ calculation.

More generally, alternative to state of the art approaches, we can employ the weak metric discussed in Sec. 2.3.4 and 2.3.6. The choice of the metric depends on our goals and tasks. Moreover, the weak distance mathematical framework is still incomplete (refer to Sec. 5.2.1).

### Hard Decision

In the case of "hard" matching decision (i.e. finding only one best match), we employ the Maximum A Posteriori (MAP) classification or risk minimization with some cost function (where $E$(costs) = risks). By MAP classification, the

$P(c|r)$ is interpreted as a posterior DoB for the matching and is performed as follows:

$$\hat{c}_{MAP} := \arg\max_{c \in \{C\}_r} P(c|r), \tag{3.24}$$

where $\{C\}_r$ denotes all selected by tree search nodes.

The risk minimization employment implies a search of the optimum decision with some function $l(\hat{c},c)$ that describes costs of the $\hat{c}$ decision when the true concept is $c$:

$$\hat{c}_l := \arg\min \sum_{c^* \in \{C\}_r} l(c^*,c) P(c|r). \tag{3.25}$$

### 3.5.3 Concepts Learning

In [Bel12d] we justify the necessity of the concepts learning for intelligent autonomous systems:

> To cope with open-world modeling, an autonomous system must be able to extend its knowledge beyond prior information by acquiring new concept definitions. Another purpose of knowledge is thus to store information learned from experience during operation. The task of dynamically expanding the knowledge of an autonomous system constitutes an extension to the classical concept learning problem.

An autonomous system can extend its prior knowledge by learning specific entities as well as class concepts. While the former is straightforward, the latter is not clear due to ambiguous choice of the structure of attributes and values clustering. The first consideration can be the following: representatives matched to dummy concepts (Fig. 3.7b) are candidates for class concept learning. The second consideration can be the *minimum description length* (MDL) [Grü07] – a description length $L_D$ of a world model

$M$ given a new concept $C$ added to a weighted description length of this concept has to be minimal sum over time span $[t_0; t_n]$:

$$L_{D,min} = \arg\min\left(\sum_{t=t_0}^{t_n} L_D(M(t)|C(t_0))) + wL_D(C(t_0))\right). \qquad (3.26)$$

For a more detailed discussion of the MDL approach in concepts learning refer to [Kuw13].

# 4

# Experiments and Evaluation

## 4.1 World Modeling Realization

In order to experimentally test presented concepts and methods, we have developed a universal world modeling subsystem, called *Progressive Mapping with Prior knowledge Matching* ($PM^2$). It incorporates the following components:

- tracking modules;
- dynamic modeling with Progressive Mapping framework;
- qualitative and quantitative information assessment modules;
- prior knowledge ontology system;
- information matching and interchange mechanisms;
- 3D scene and attributes visualization modules;
- network-based distributed infrastructure.

The system was deployed and tested in ARMAR-III robots [Deu08]. Some parts were implemented and tested within the NEST project [Bau09b].

## 4.2 Modeling Architecture and Realization

*Workflow* The overall world modeling $PM^2$ architecture is presented in Fig. 4.1. The implied workflow is as follows:

1. prior knowledge definition – before autonomous system operation starts, a group of experts defines prior knowledge relevant to the world of interest;

2. information acquisition – sensors subsystem observes the world of interest, pre-processes data, extracts relevant information about the environment and sends it over network to the dynamic modeling subsystem;

3. dynamic modeling – dynamic modeling modules process the incoming information by progressing information containers, matching prior knowledge, etc.;

4. prior knowledge matching – prior knowledge modeling subsystem processes requests from dynamic modeling subsystem for information matching, complementation and learning;

5. monitoring – monitors visualize contents of the modeling subsystem.

*Sensors Subsystem* The sensor subsystem complies intrinsic knowledge restrictions, performs signals pre-processing and delivers relevant information in form of zip-compressed XML blocks over network.

*Network* The network infrastructure is represented by client-server connections over TCP/IP. Such an approach allows for a distributed and asynchronous information interchange.

*Dynamic Modeling* The dynamic modeling subsystem has to fulfill a variety of challenging tasks (refer to Sec. 2.2.2). With this goal in mind, we have developed a distributed, modular, multi-threading cross-platform subsystem for models management and information fusion. In practice, this

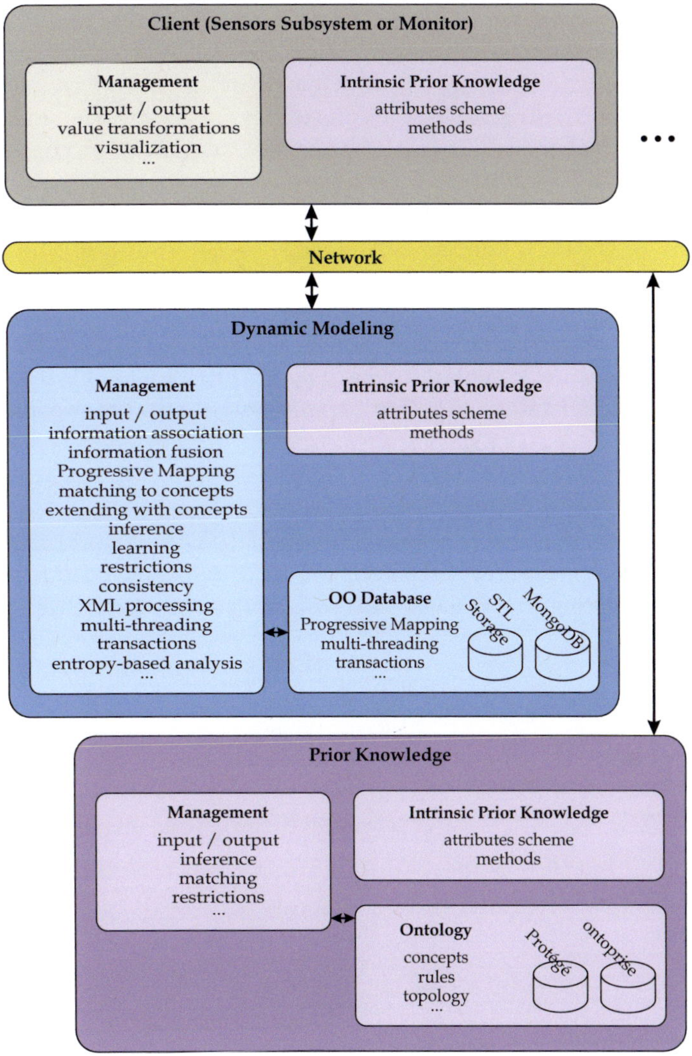


**Figure 4.1:** Architecture of the object-oriented world modeling system $PM^2$.

subsystem is written in C++ with STL/Boost coupling and represents an object-oriented database (configurable to employ either a proprietary in-memory database or mongoDB) with managers for network connections, serializations, information fusion, consistency checks, etc. The realization for NEST is implemented in Java with Apache Commons Math coupling.

*Prior Knowledge* Prior knowledge contains expert knowledge about the world of interest (e.g. concepts, rules, etc.) as well as intrinsic prior knowledge (e.g. global scheme of attributes, as described in Sec. 1.3.3 and Sec. 2.2.1). Prior knowledge subsystem supports concepts matching and complementation of representatives. Practically, we have implemented the prior knowledge subsystem as Protégé ontology (Fig. 4.2) but it is possible to use any other ontology database (e.g. Ontoprise). For test purposes, we have predefined knowledge for a kitchen environment (class and entity concepts of plates, cups, tables, etc.).

*Monitors* Monitor clients display contents of the current world model. To this end, remote monitors keep a copy of the dynamic model locally (with constant updating) and visualize a 3D scene with open-source OGRE library.

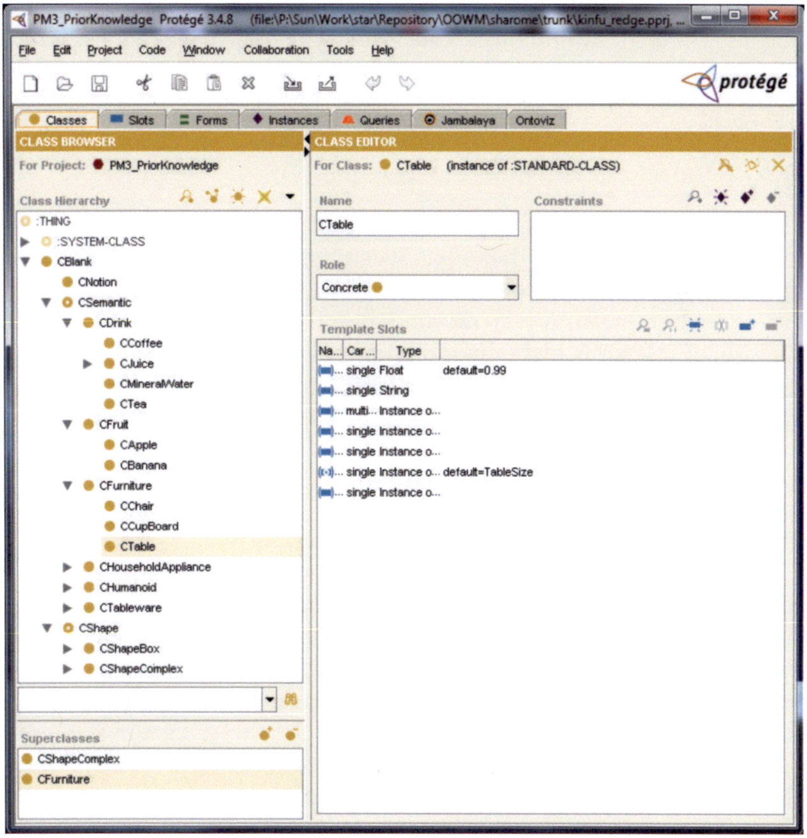


**Figure 4.2:** Protégé ontology framework: modeling of concepts and their hierarchies.

## 4.3 Test Scenarios and Demonstrators

The presented research and developments can be used with arbitrary sensors and autonomous system set-ups. In this work, the developed system and a selection of sub-modules were tested with three hardware platforms: humanoid robot head ARMAR-III [Asf08] (Fig. 4.5a), complete robot ARMAR-III [Asf06] (Fig. 1.8) and surveillance system NEST [Bau09b] (Fig. 1.7).

### 4.3.1 Robotic Set-Up at Fraunhofer IOSB

The IOSB robot head was a part of ARMAR-III project that is discussed in the next section. The head has seven positional degree of freedoms and is equipped with two stereo cameras (focal length of 6 and 8 mm) and a microphone array with six condenser Lavaliere microphones. Since the overall system is highly modular and distributed, it can operate on a computer cluster if more computation power is needed (when e.g. the system is deployed in the open world instead of a laboratory environment[1].

### 4.3.2 ARMAR Robot at KIT

The Collaborative Research Center 588 "Humanoid Robots – Learning and Cooperating Multimodal Robots" (SFB 588) was established on the 1st of July, 2001 by the Deutsche Forschungsgemeinschaft (DFG) and had been running until the 30th of June, 2012. The center is assigned to the Department of Computer Science of the Karlsruhe Institute of Technology. During the third phase of the project (2008-2012 years), complex world modeling, exploration, intelligent perception, speech processing and actuation concepts were investigated and experimentally proved. Over its lifetime, the project involved more than 90 scientists and about 13 institutes, who produced more than 450 scientific publications. The main research and development platform of the Center was the ARMAR robot. Four generations of the platform have been developed (Fig. 4.3).

1 This can, for example, make auto-calibration or information association and classification much more complicated, if vast sets of concepts and representatives are required.

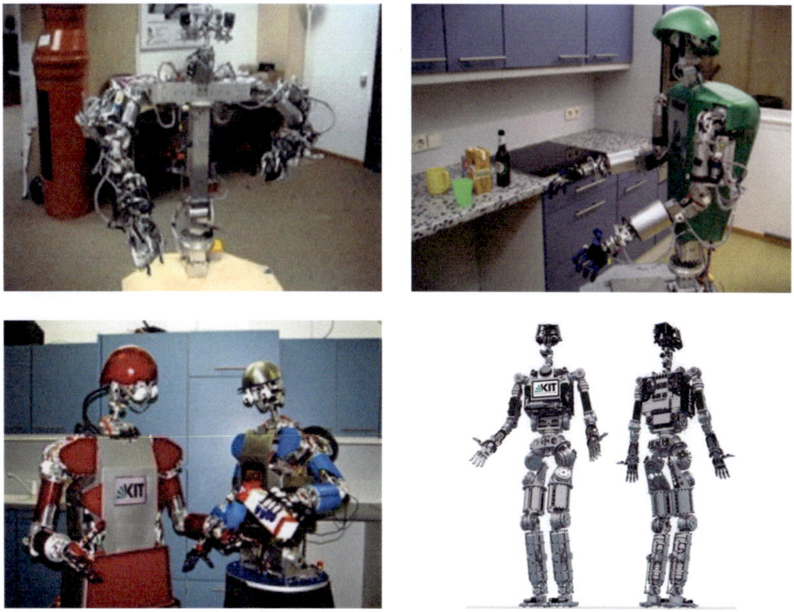

**Figure 4.3:** The four generations of the humanoid robot ARMAR ([Deu08]) placed in reading order: starting at the top left position with the first generation and ending at the bottom right position with the fourth generation.

The main experimental platform used during the present analysis is the ARMAR robot of the third generation (ARMAR-III) (Fig. 1.8). It consists of above mentioned robot head, 7 degrees of freedom (DoF) arms (position, velocity and torque sensors; 6D FT-sensors, sensitive skin), 8 DoF hands (pneumatic actuators, holding force 2.5 kg), 3 DoF torso (2 embedded PCs, 10 DSP/FPGA units) and holonomic mobile platform (3 embedded PCs, 3 laser scanners, 2 car accumulators for 2-3 hours of autonomous operation) [Asf06].

### 4.3.3 NEST at Fraunhofer IOSB

The NEST platform developed in Fraunhofer IOSB is a next generation automated surveillance and monitoring system. It implements decentralized, service-oriented system architecture with intelligent information processing, including motion detection and tracking, semantic description of complex situations, etc. [Moß10]. The implemented mechanisms allow for detection of e.g. abnormal activity or abandoned entities and present the complete state estimate history of selected persons. As a testing installation, NEST operates a network of multi-modal sensors within the IOSB building and at the surrounding grounds.

## 4.4 Evaluation

### 4.4.1 Knowledge-Driven Scene Analysis and Interactive Exploration

During a cooperation with SFB 588 work-groups concerned with perception (including collaborative Arbeitsgruppe 1 "Perception and World Modeling"), we have performed a set of experiments in order to test the possibility of interactive exploration of unknown entities, supported by a knowledge-driven scene analysis. At first, the ARMAR robot performs environment exploration, i.e. world of interest (WoI) information acquisition by active observation (e.g. change of sensors' position and direction) or active change of the WoI (e.g. taking things into hands or moving things around). Further, the robots starts to explore the environment interactively: it communicates with humans and collects the needed information. This interaction includes dialogs, gestures and tracking of entities.

In order to perform such complex tasks, the developed modeling subsystem $PM^2$ was coupled with many other subsystems (e.g. dialog modules, face recognition modules, etc.) and in the first place with an opto-acoustic scene analysis (OPASCA) [Mac10a] subsystem.

At first, all components for the hierarchical knowledge-driven opto-acoustic scene analysis were connected in a pipeline manner with the $PM^2$ at the end. This allows to shift information processing from a short-term tracking basis (memory over several frames) onto a prolonged temporal analysis. Particularly, this allows for modeling of entities, which have left the observable area but remain in the vicinity. Additionally, it allows employment of the global prior knowledge database. Finally, it allows also re-recognition of learned entities.

The first experimental results of such workflow are presented in [Küh10] with an example of a real life cycle of a representative during the environment exploration. The explored scenario was as follows: a loudly speaking person enters an observable area of the Fraunhofer IOSB robotic set-up, talks to the system and leaves the observable area silently. The Figure 4.4 represents a life cycle of the corresponding representative: at the beginning the person is detected by microphones, localized and at time step $t_0$ created within the dynamic model as a point-like representative. The incoming observations are being fused into representative's description, its existence probability and position confidence level are increasing. Upon reaching a certain threshold by the existence probability, a matching to prior knowledge is started (time step $t_1$). The TYPE attribute description reaches a certain threshold in confidence level at $t_2$ resulting in classifying the entity as a PERSON. This triggers modules for face recognition and height estimation. At the moment $t_3$, the person starts to look to the camera, which finally leads to the identification of the person (e.g. TIMO). At the time point $t_4$, the person silently leaves the observable area – the existence probability and the position confidence level starts to decrease. At last, the existence probability reaches a deletion threshold at $t_5$ and the representative is removed from the scene.

As discussed in [Küh10], the world modeling subsystem provides a consistent and efficient information storage during all phases of the scene analysis (i.e. detection/instantiation, specialization and deletion). In particular, the modeling subsystem provides functionality to store and exchange information between all sub-modules on the level of representatives and DoB distributions.

Further experiments were carried out at more complex set-ups intended for an interactive exploration. The ARMAR head and arm platform at Fraunhofer IOSB (Fig. 4.5) was challenged with the following scenario: at the beginning, the robot head observes its surroundings. As soon as a person steps into the world of interest area, the robot starts to follow the person

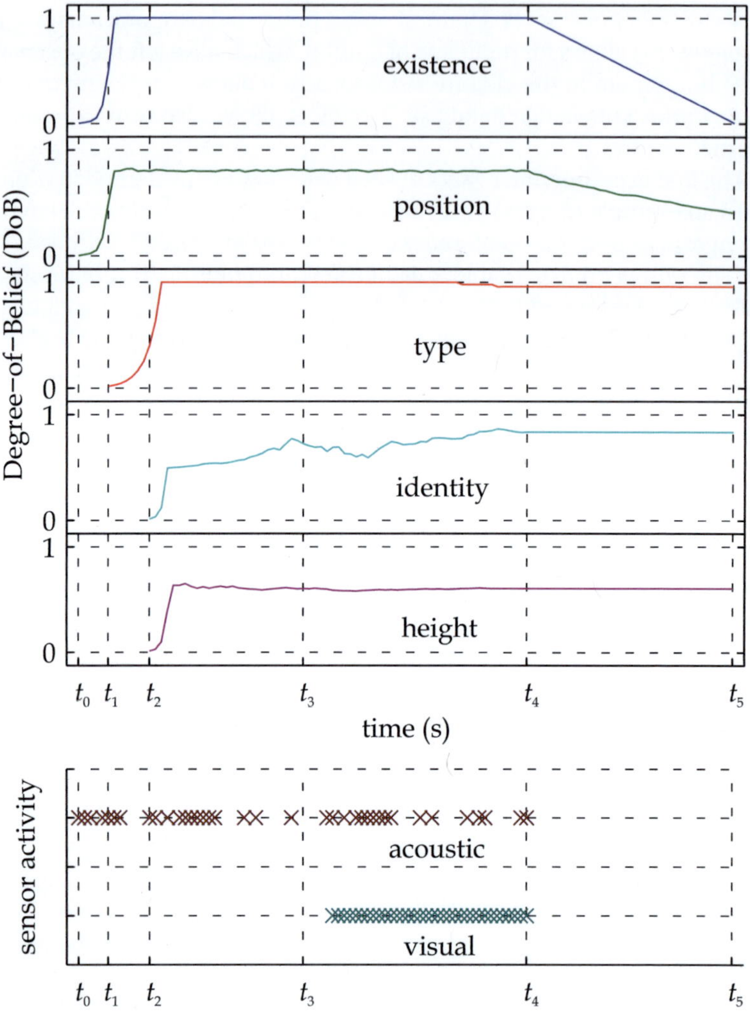


**Figure 4.4:** Life cycle of a representative during the knowledge-driven opto-acoustic scene analysis [Küh10].

with cameras and head turns. The head tries to recognize the person and personally greets him or her upon a successful recognition. By a speech command from the person, the robot head starts to examine a table in front of it. It tries to recognize known entities and asks about the unknown ones, extending its dynamic model and prior knowledge. After all desired entities were labeled, the robot plays a memory game: it turns for a while away, while the person removes some entities from the table, adds some new and moves some to a new place. Hereafter, the robot head turns back and names the changes. Upon human request, the robot can point requested entities with its hand. Moreover, during the whole episode, any loud sound (e.g. of a fallen down entity) can attract robot's attention and switch to environment observation of a corresponding area of attraction.

During this complex episode, many involved advanced methods and concepts of the humanoid robotics have to work together within more or less real environment:

- geometrical auto-calibration, color and light balance;
- entities detection, segmentation, recognition, learning;
- person detection, tracking, recognition;
- speech recognition, dialog manager;
- gestures and their direction recognition;
- dynamic modeling;
- prior knowledge matching and update;
- etc.

In this scenario, the world modeling subsystem served as a central information hub for all other subsystems. Thereby, the prior knowledge was pre-defined as follows:

1. information for persons recognition: 5 persons (opto-acoustical patterns [Swe10] of faces, heights and speech features);

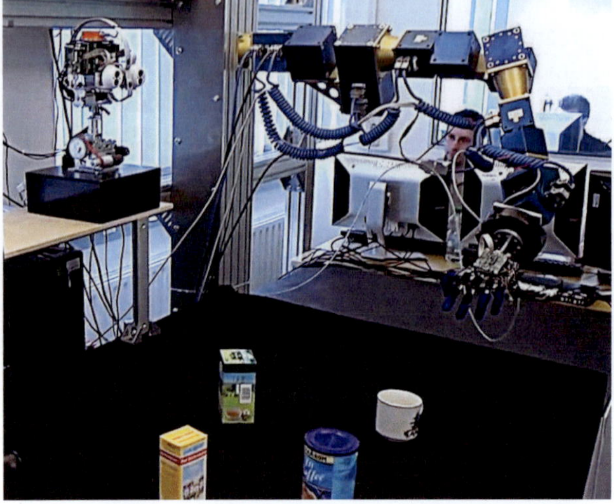

**(a)** Robot head and arm platform at Fraunhofer IOSB;

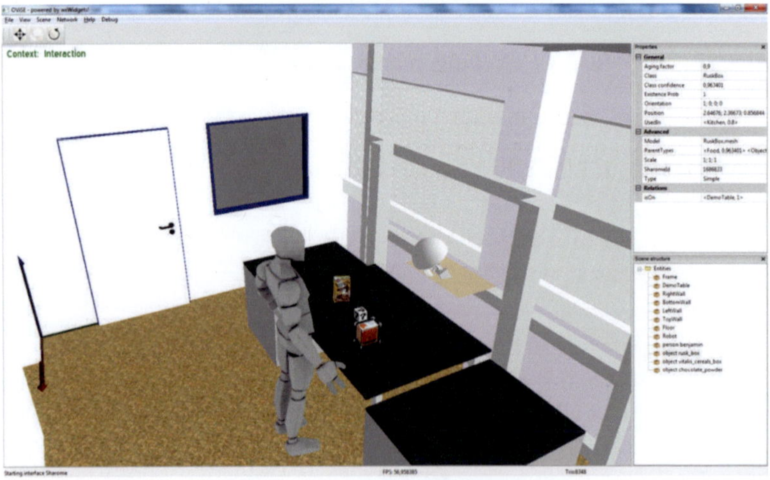


**(b)** Visualization of the dynamic 3D model;

**Figure 4.5:** Interactive exploration episode.

2. information for things recognition: 20 entities (opto-acoustical patterns [Swe10] for forms, sizes, textures and sounds; created by 3D laser scanner and camera (Fig. 4.6a) precise models and textures (Fig. 4.6b), attributes defined in Protégé ontology);

3. information for type classification: 30 classes (concepts and their relations defined in Protégé ontology);

4. information for hierarchical type search and information extension: 2 hierarchies (form and semantic);

5. topology information: 3D laboratory model based on manual measurements.

**(a)** 3D laser scanner and textures camera; **(b)** Models of scanned entities;

**Figure 4.6:** Experimental set-up for complex objects modeling and model examples [Deu08].

The exploration and the analysis ran in real time on two Intel Core 2 Quad 2.67 GHz PCs (one for OPASCA and one for $PM^2$). With described above conditions, the number of mismatching during recognition was around 7-12% based on approximately 50 tries with multiple (4-6) entities each. However, the exact exploration and recognition correctness of the workflow is hard to assess due to following issues:

- the uncertainty of the sensors is relatively large and depends on the direction and the distance to entities;

- attribute descriptions are limited to sizes, color patterns and sounds;
- entities of similar sizes and color patterns are mismatched so often, that we had to influence on the choice of entities (maximum color and size separation). This could naturally introduce bias.

More realistic scenarios would be even harder to assess due to further systematic uncertainties, i.e. difference in light conditions (sunlight through windows, diffuse artificial light, directed desk light), position and direction auto-calibration issues by head movements (pitch, yaw, roll), etc. In spite of white balance and brightness auto-leveling, mismatching dominates even with relatively small (i.e. 8-10) entities sets. In order to reduce bias, the lighting condition had been keeping constant (laboratory environment). The position and direction re-calibration was improved when the head position was fixed and extra markers were placed on the laboratory walls.

As a next experiment, the environment observation in a designed interactive manner was performed as a background process on the ARMAR-III robot during its normal operation. The prior knowledge was extended and refined by the following information:

1. topology information: precise 3D kitchen model (Fig. 4.7b) based on laser measurements (Fig. 4.7a) and camera observations;
2. complex entities information: generic knowledge about cupboards, dishwashers, etc. (e.g. doors, drawers, hinges and joints);
3. statistical scene analysis information: probabilistic position distributions for entities (i.e. usual placements of things within the modeling environment, e.g. cups on the cupboard or on the table).

### 4.4.2 Bayesian Filtering

In order to experiment with Bayesian Filtering (refer to Sec. 2.3), we have implemented two state of the art filters – the Kalman and particle filters – within the NEST project. Since the NEST project is dedicated to monitoring and surveillance of indoor and outdoor areas, the modeling domain has to include area topology, available sensors, targets and immaterial (e.g. events)

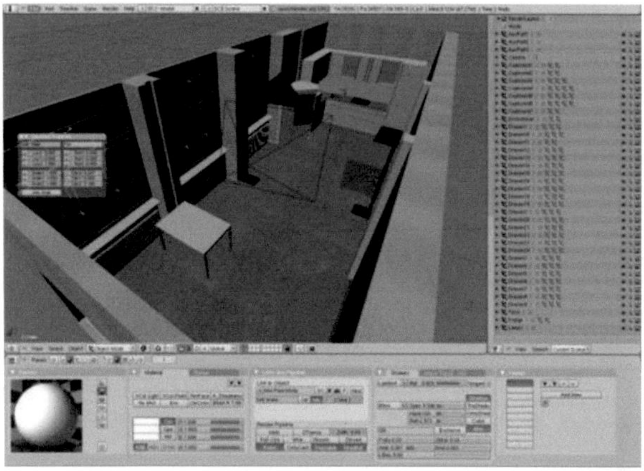

**(a)** Raw kitchen room data: triangulated surfaces as a result of laser measurements;

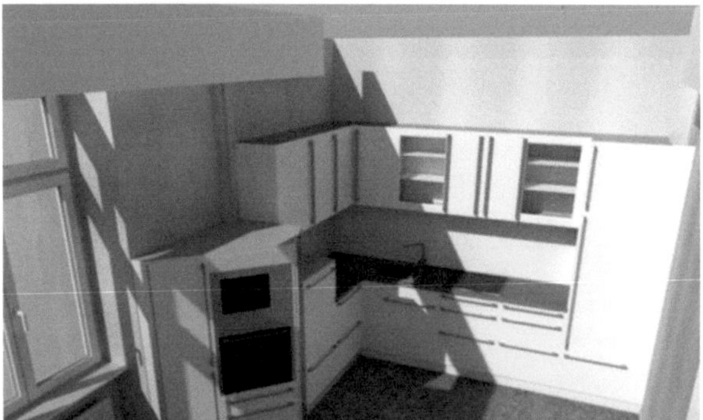

**(b)** 3D kitchen model after manual processing of the raw data (combining triangles into flat surfaces, segmenting and semantic labeling of representatives, etc.);

**Figure 4.7:** Kitchen model as a prior knowledge for the world of interest [Deu08].

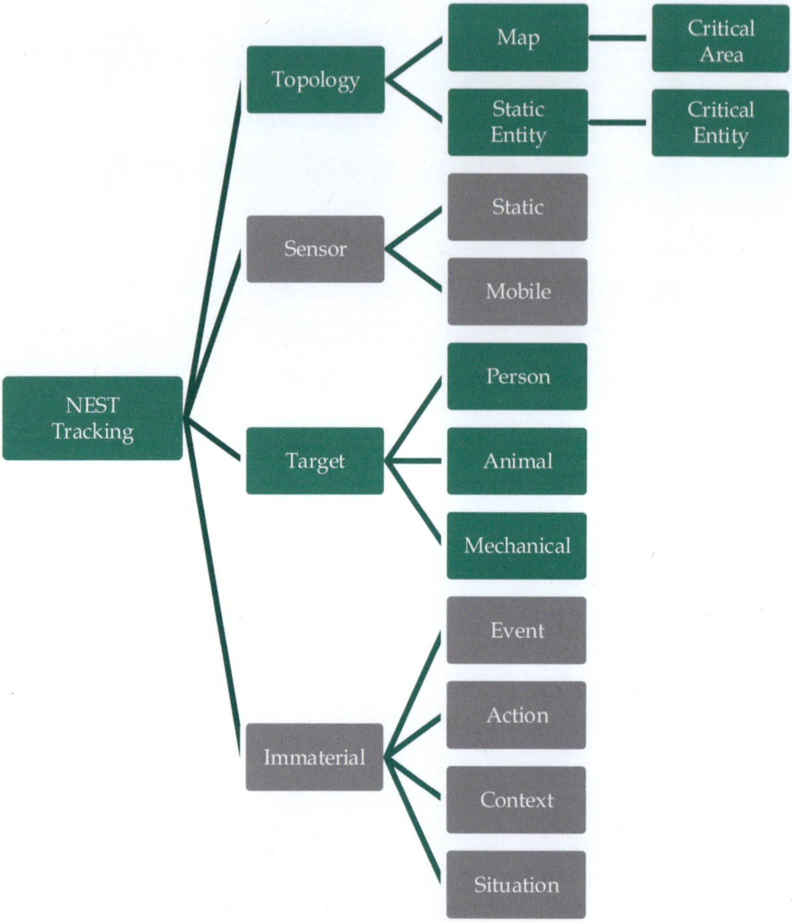


**Figure 4.8:** NEST modeling domain hierarchy.

entities. The employed domain hierarchy is presented in Fig. 4.8. Hereby, the domain representatives contain:

- Topology: type, state, coordinates, direction, bounding box, covariance, relation, critical mark;
- Sensor: type, state, coordinates, direction, covariance;
- Target: type, state, coordinates, speed, direction (e.g. face direction), bounding box, covariance, color (histogram), relations;
- Immaterial: type, coordinates, direction, covariance, relations, extended information.

### Kalman Filter 2D Tracking

Let us apply a Kalman filter (refer to Sec. 2.3.2) to tracking of a person on a street. Here we are interested in position $[x,y]^T$ and velocity $[\dot{x},\dot{y}]^T$ estimation – so, the state vector $\vec{s}$ is:

$$\vec{s} = \begin{bmatrix} x \\ \dot{x} \\ y \\ \dot{y} \end{bmatrix}. \tag{4.1}$$

Since both position and velocity are state components, they will not be overwritten by current sensor values or calculated in the manner of $\dot{x} = (x_k - x_{k-1})/\Delta t$ but will be updated by (and so will take advantages of) the inference process.

The speed of the person is assumed to be constant (i.e. *constant velocity model*) by two reasons: it simplifies the estimation and it is often close to real life behavior (if a person is moving with 5 km/h velocity, in the next moment he or she will probably have approximately the same speed). For this reason, the evolution matrix $\boldsymbol{F}$ is constant.

A real person, however, can change his or her own speed due to many reasons: because of passing by of other pedestrians, obeying traffic lights, following the intended path plan, etc. We reflect it by defining a process noise $\vec{w}$ as acceleration $\vec{a}_r \in N(0,\sigma_{a_r})$. We also specify probability distributions of acceleration values by normal distributions. This reflects the fact that in most of the cases the acceleration is close to 0 (soft speed-up or slow-down). The width of the allowed acceleration distribution is specified by an empirical constant $\boldsymbol{\sigma}_{a_r}$ that is assigned proportional to physical limitations (e.g. person's acceleration). Thus, the process covariance matrix $\boldsymbol{Q}$ is also constant.

We make observations every $\Delta t$ seconds. Hereby, this time interval defines the temporal discretization step $(k-1) \xrightarrow{\Delta t} k$ for system changes. In the time periods between observations, the acceleration $\vec{\boldsymbol{a}}_r$ is supposed to be constant.

As soon as sensors detect a person, the initial state can be formed according to the observation data as well as the covariance matrix initialized by a sufficiently large number $L$:

$$\begin{aligned} \vec{\boldsymbol{s}}_{0|0} &= \begin{bmatrix} x_{\text{obs}} \\ 0 \\ y_{\text{obs}} \\ 0 \end{bmatrix}, \\ \boldsymbol{P}_{0|0} &= \begin{bmatrix} L & 0 & 0 & 0 \\ 0 & L & 0 & 0 \\ 0 & 0 & L & 0 \\ 0 & 0 & 0 & L \end{bmatrix}. \end{aligned} \tag{4.2}$$

The evolution of the true state is given by 2.13:

$$\tilde{\vec{\boldsymbol{s}}}_k = \boldsymbol{F}_k \tilde{\vec{\boldsymbol{s}}}_{k-1} + \underbrace{\boldsymbol{B}_k \vec{\boldsymbol{u}}_k}_{0} + \underbrace{\vec{\boldsymbol{w}}_k}_{\boldsymbol{G}\vec{\boldsymbol{a}}_r}, \tag{4.3}$$

where we (observers) do not have any influence on the person, so, $\boldsymbol{B}_k \vec{\boldsymbol{u}}_k = 0$, the process noise is defined by $\vec{\boldsymbol{a}}_r$, called *random disturbance source* or *random acceleration,* and *disturbance transfer matrix* **G**, which defines a transfer of the random disturbance onto each state parameter (i.e. how much this acceleration will affect position and speed of the person till the next time step).

Since we imply constant velocity model, the evolution matrix shifts the position according to speed and keeps the speed constant:

$$\begin{cases} x_k = x_{k-1} + \dot{x}_{k-1}\Delta t \\ \dot{x}_k = \dot{x}_{k-1} \\ \text{similar for } y \end{cases} \Rightarrow \boldsymbol{F} = \begin{bmatrix} 1 & \Delta t & 0 & 0 \\ 0 & 1 & 0 & 0 \\ 0 & 0 & 1 & \Delta t \\ 0 & 0 & 0 & 1 \end{bmatrix}. \tag{4.4}$$

The acceleration, disturbance on position and speed, and disturbance transfer matrix are given by:

$$\vec{\boldsymbol{a}}_r = \begin{bmatrix} \ddot{x} \\ \ddot{y} \end{bmatrix}, \tag{4.5}$$

$$\begin{cases} \Delta x = \frac{\ddot{x}\Delta t^2}{2} \\ \Delta \dot{x} = \ddot{x}\Delta t \end{cases} \Rightarrow \boldsymbol{G} = \begin{bmatrix} \Delta t^2/2 & 0 \\ \Delta t & 0 \\ 0 & \Delta t^2/2 \\ 0 & \Delta t \end{bmatrix}. \tag{4.6}$$

The process covariance matrix $\boldsymbol{Q}$ is defined by noise transfer matrix $\boldsymbol{G}$ and acceleration covariance matrix $\boldsymbol{\sigma}_{a_r}$:

$$\boldsymbol{\sigma}_{a_r} = \begin{bmatrix} \sigma_{\ddot{x}}^2 & 0 \\ 0 & \sigma_{\ddot{y}}^2 \end{bmatrix}, \tag{4.7}$$

$$\boldsymbol{Q} = \boldsymbol{G}\boldsymbol{\sigma}_{a_r}\boldsymbol{G}^{\boldsymbol{T}} = \begin{bmatrix} \frac{\Delta t^4 \sigma_{\ddot{x}}^2}{4} & \frac{\Delta t^3 \sigma_{\ddot{x}}^2}{2} & 0 & 0 \\ \frac{\Delta t^3 \sigma_{\ddot{x}}^2}{2} & \Delta t^2 \sigma_{\ddot{x}}^2 & 0 & 0 \\ 0 & 0 & \frac{\Delta t^4 \sigma_{\ddot{y}}^2}{4} & \frac{\Delta t^3 \sigma_{\ddot{y}}^2}{2} \\ 0 & 0 & \frac{\Delta t^3 \sigma_{\ddot{y}}^2}{2} & \Delta t^2 \sigma_{\ddot{y}}^2 \end{bmatrix}. \tag{4.8}$$

In the case of active prior knowledge employment, we can adjust matrices $\boldsymbol{\sigma}_{a_r}$, $\boldsymbol{F}$ and $\boldsymbol{B}$ on-the-fly in order to reflect road topology, traffic lights and obstacles, as well as initiate direct input (e.g. by sending a police car to stop the target) that is currently 0. Additionally it is possible to calculate $\boldsymbol{Q}_k$ and $\boldsymbol{R}_k$ from incoming data with *Autocovariance Least-Squares* (ALS) [Ode06b] method.

Finally, the true state evolution from Eq. 4.3 looks like:

$$\begin{bmatrix} x \\ \dot{x} \\ y \\ \dot{y} \end{bmatrix}_k = \underbrace{\begin{bmatrix} 1 & \Delta t & 0 & 0 \\ 0 & 1 & 0 & 0 \\ 0 & 0 & 1 & \Delta t \\ 0 & 0 & 0 & 1 \end{bmatrix} \begin{bmatrix} x \\ \dot{x} \\ y \\ \dot{y} \end{bmatrix}_{\text{true},k-1}}_{\substack{\text{motion with constant speed (propagation)} \\ \begin{bmatrix} x+\dot{x}\Delta t \\ \dot{x} \\ y+\dot{y}\Delta t \\ \dot{y} \end{bmatrix}_{\text{true},k-1}}} + \underbrace{\begin{bmatrix} \Delta t^2/2 & 0 \\ \Delta t & 0 \\ 0 & \Delta t^2/2 \\ 0 & \Delta t \end{bmatrix} \begin{bmatrix} \ddot{x} \\ \ddot{y} \end{bmatrix}_k}_{\substack{\text{random acceleration contribution (noise)} \\ \begin{bmatrix} \ddot{x}\Delta t^2/2 \\ \ddot{x}\Delta t \\ \ddot{y}\Delta t^2/2 \\ \ddot{y}\Delta t \end{bmatrix}_k}} . \tag{4.9}$$

The measurement $\tilde{\vec{o}}$ is given by 2.14:

$$\tilde{\vec{o}}_k = \boldsymbol{H}_k \tilde{\vec{s}}_k + \vec{v}_k. \tag{4.10}$$

The sensor parameters are supposed to be constant, leading to $\boldsymbol{H}, \boldsymbol{R} = const$. Since our camera can see only a position of the person, the observation model matrix is defined as:

$$\boldsymbol{H} = \begin{bmatrix} 1 & 0 & 0 & 0 \\ 0 & 0 & 1 & 0 \end{bmatrix}. \tag{4.11}$$

The measurement noise is assumed to be $\vec{v} \in N(0, R)$. We assume the following form of the observation covariance matrix:

$$\boldsymbol{R} = \begin{bmatrix} \sigma^2_{\text{obs},x} & 0 \\ 0 & \sigma^2_{\text{obs},y} \end{bmatrix}. \tag{4.12}$$

Finally, the measurement in Eq. 4.10 is now:

$$\begin{bmatrix} x \\ y \end{bmatrix}_{obs,k} = \underbrace{\begin{bmatrix} 1 & 0 & 0 & 0 \\ 0 & 0 & 1 & 0 \end{bmatrix}_k \begin{bmatrix} x \\ \dot{x} \\ y \\ \dot{y} \end{bmatrix}_{\text{true},k}}_{\substack{\text{ideal observation without noise} \\ \begin{bmatrix} x \\ y \end{bmatrix}_{\text{true},k}}} + \underbrace{\vec{v}_k}_{\substack{\text{clutter (e.g. sensor noise)} \\ \begin{bmatrix} \propto \sigma_{\text{obs},x} \\ \propto \sigma_{\text{obs},y} \end{bmatrix}_k}} . \tag{4.13}$$

The proposed Kalman filter parametrization was tested in NEST project with resulting experimental samples presented in Figure 4.9.

### Particle Filter 2D Tracking

The NEST system is capable of tracking entities (e.g. people, cars, ships) in closed (e.g. buildings) and open (e.g. seas) areas. In practice we have implemented a particle filter (refer to Sec. 2.3.5) with the following workflow.

At the time point $k$ we have detected a person. The position attribute $X$ of the entity is initialized by a continuous multivariate Gaussian distribution $p(\vec{x})$ for coordinates and speed. Since the estimate converges within a few steps, we initialize this distribution with a large variance (at least of $3\sigma$ of sensors' resolution) and assume the speed parameter to be about $\pm 12\text{m}/s$ (unknown speed, so maximum possible value; Fig. 4.10a-4.10b). At this time step, we draw $P$ uniform (equally weighted) particles from this initial distribution. The positions of particles are chosen according to $p(\vec{x})$. In order to visualize the state estimate, we can convert particles into Gaussian mixture with the following procedure: particle position → mean, LDQ → covariance matrix. Upon next observation (time step $k+1$), we propagate particles according to our system dynamic equations. Since the speed is unknown, the particles are scattered in all directions on the coordinate plane (Fig. 4.10a-4.10b). Next, we fuse new observation into obtained prediction. To this end, we estimate sensor parameters at the observed values and calculate likelihood (Fig. 4.10e). Then, we reweight particles accord-

(a) Entity detection in video streams [Bel12d];

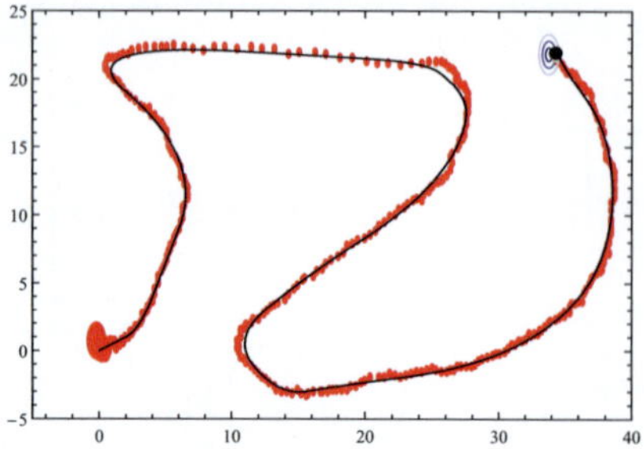


(b) Simulation: ground truth (black), estimate (red), observation (blue);

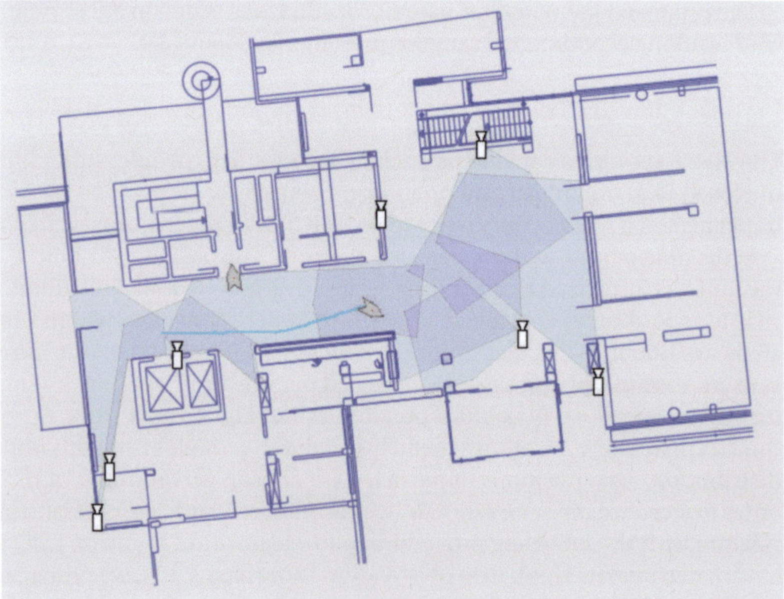

(c) World model with topology and observed persons in indoor surveillance [Bel12d];

**Figure 4.9:** Kalman filter tracking in NEST project.

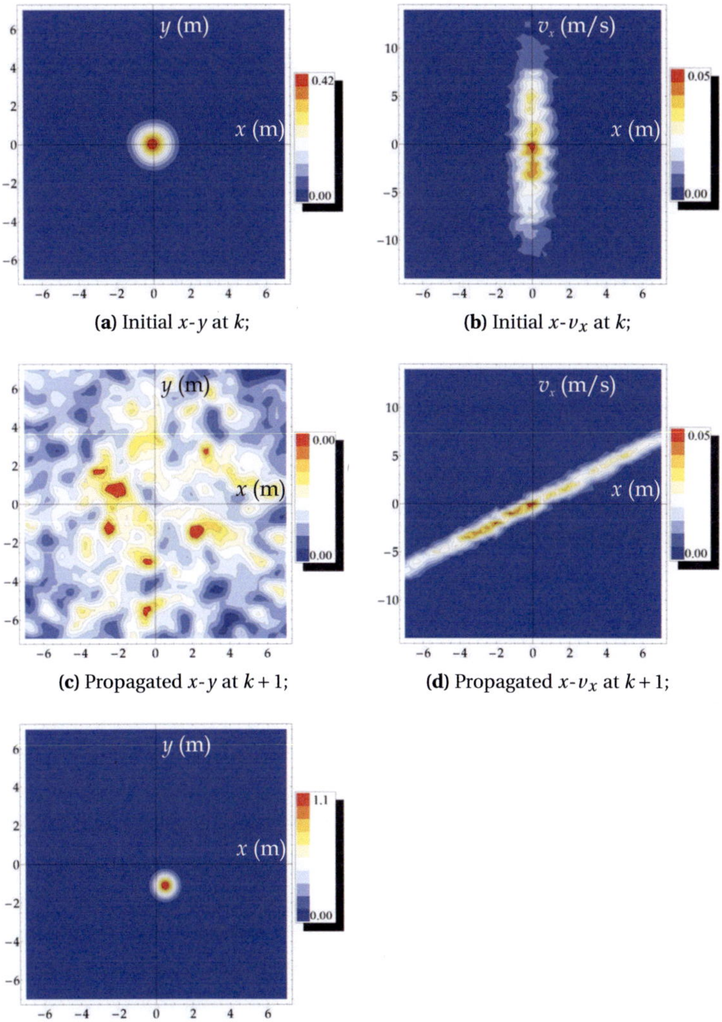


**(a)** Initial $x$-$y$ at $k$;

**(b)** Initial $x$-$v_x$ at $k$;

**(c)** Propagated $x$-$y$ at $k+1$;

**(d)** Propagated $x$-$v_x$ at $k+1$;

**(e)** Likelihood $x$-$y$ at $k+1$;

**Figure 4.10:** Slices of the multivariate state estimate p.d.f.

ing to likelihood function (Fig. 4.12a). Slices of a reweighted estimate are presented in Fig. 4.11a-4.11b. It is quite possible that only a few particles

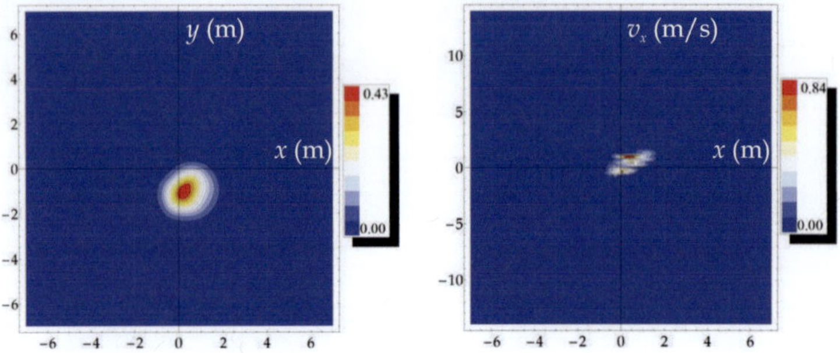


**(a)** Reweighted $x$-$y$ at $k+1$;

**(b)** Reweighted $x$-$v_x$ at $k+1$;

**Figure 4.11:** Slices of the reweighted multivariate state estimate p.d.f.

have obtained a significant weight since, for example, likelihood is shifted relative to most of particles. This is visible, for example, in Fig. 4.11b. Such process of degeneracy (i.e. loosing particles) demands resampling in order to maintain more or less the same number of particles along the iterations. For resampling, we converse particles into GM, based on particle positions and LDQs, and sample it (Fig. 4.12b). A resulting set of new particles is assumed to be a prior distribution for the next step propagation for the moment $k+2$.

### Sampling with Weak Metric Algorithm

We have discussed the weak metric approach related to sampling in [Pak13], as well as in Sec. 2.3.4 and 2.3.6. For illustration, let us consider sampling of a 40-component Gaussian mixture. In order to compare our method with state of the art approaches, we have generated 100 uniformly weighted particles according to the distribution density and the same amount of samples resulting from the WD algorithm (optimized for position and weights). For the latter, we have employed 200 Gaussian probe functions similar to those described in Sec. 4.4.3. The results are presented in Fig. 4.13 and Table 4.1.

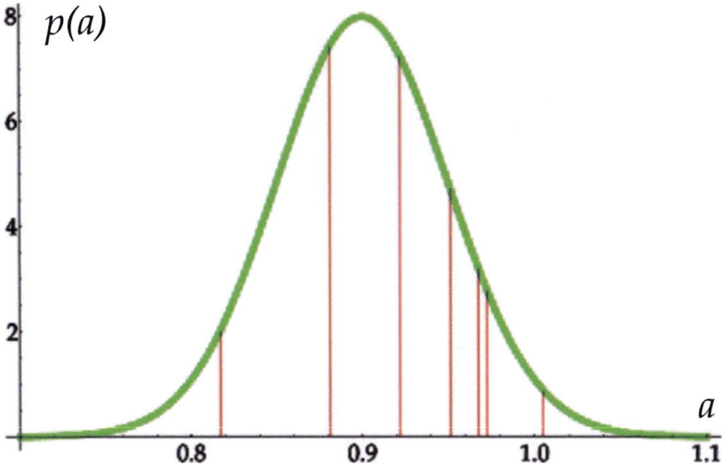


**(a)** Particles reweighting: green – likelihood, red – weights;

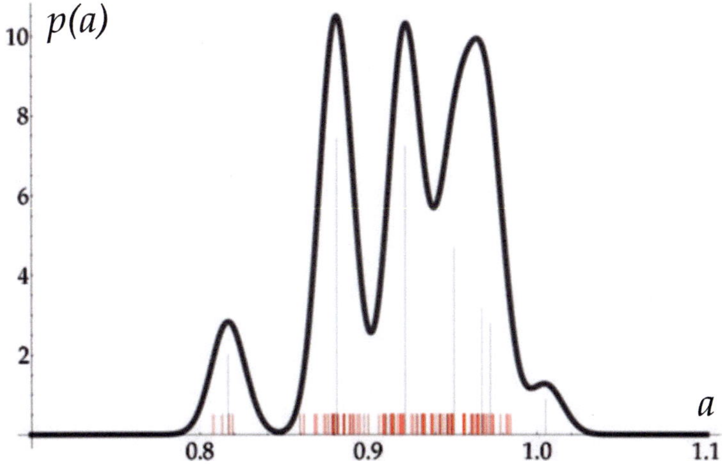


**(b)** Resampled state estimate: gray – reweighted according to likelihood particles from (a), thick black – resulting GM after particles conversion, red – new particles after resampling;

**Figure 4.12:** Resampling: reweighting and conversion of particles to GM.

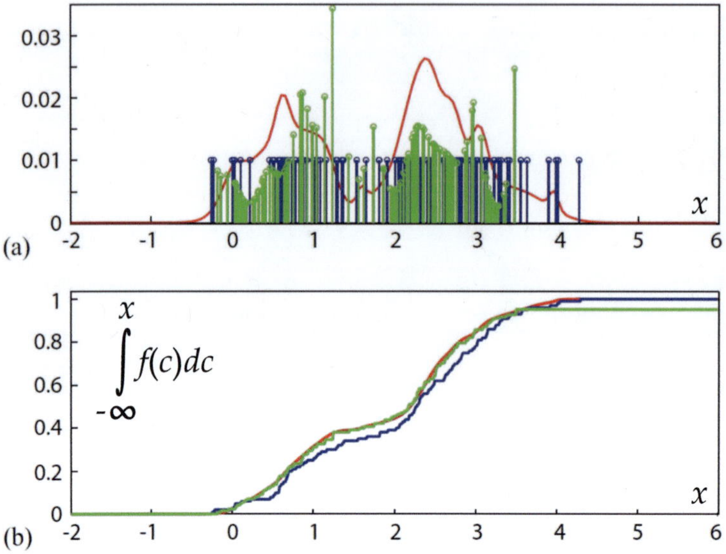


**Figure 4.13:** Gaussian mixture sampling [Pak13]: (a) original GM (red), 100 uniformly weighted samples (blue), and 100 samples resulting from the WD algorithm (green). Line heights represent relative weights; (b) Cumulative distributions of the same data as in (a). The WD cumulative distribution (green) is equal to the original one (red) till the right part, which is affected by window function $G(\vec{\boldsymbol{\omega}}, \boldsymbol{W}; \vec{\boldsymbol{\mu}})$ suppression during the modulation process.

| Method | WSD | Run-time [s] | Mean | Variance | Norm |
|---|---|---|---|---|---|
| Original | - | - | 1.855 | 1.224 | 1.0 |
| Uniform | 0.1351 | 0.014 | 2.029 | 1.246 | 1.0 |
| WD | 0.0225 | 49.31 | 1.775 | 1.061 | 0.952 |

**Table 4.1:** A comparison of GM sampling methods [Pak13]. The gradient descent is performed over all 100 samples leading to a long run-time but has a better weak squared difference (WSD, refer to Eq. 2.31 and Eq. 2.38). Since the WD algorithm is optimized for the WSD only, the mean, variance and norm values do not follow the original values as, for example, in uniform case.

#### Extended Attributes Set

As we have discussed in Sec. 2.3.3, tracking of speed and coordinates allows basic target tracing. In more complex cases, e.g. intersecting target paths, a matching of observations to targets can be ambiguous. Hereby, it is advantageous to employ more descriptive attributes into state estimate. This allows for better observations assignment as well as better classification. One of the natural candidates for additional attributes is the color. In NEST tracking we have employed a color histogram of 10 bins normalized over all pixels in the frame. In order to test the utility of this improvement, we have performed a tracking simulation of two persons walking side by side while constantly crossing their paths (Fig. 4.14). The corresponding results are presented in Fig. 4.15. As we can see a particle filter completely fails in the case of spatial ambiguity, which is resolved by color information consideration in observation-to-estimate association.

#### Bayesian Fusion in Practice

As mentioned earlier, we modify the weights of the particles according to the likelihood. Hereby, we can also employ prior knowledge, e.g. topology restrictions (Fig. 4.16a) or state estimate (Fig. 4.16b). In Fig. 4.17 there is a result of Bayesian fusion of prior knowledge and likelihood or state estimate. The prior knowledge can be a pre-defined expert knowledge or information obtained during a statistical scene analysis [Kas13].

### 4.4.3 Gaussian Mixture Reduction Problem

We have discussed the GM reduction problematic in [Pak13], as well as in Sec. 2.3.4 and 2.3.6. For experimental tests, we have examined various reduction algorithms compared to our proposal. For example, Fig. 4.18 depicts a 1D 40-kernels GM (marked as "Original") reduced to a 8-kernels one. The "Original" distribution contains components of random mean and standard deviation $\sigma \in [0.05; 0.35]$. Hereby, we compare our WD and WC algorithms (refer to Sec. 2.3.6) to state of the art algorithms: Salmond [Pao94], West [Wes93], Runnalls [Run07], and Williams [Wil03a] ("initialization" algorithm without optimization stage). The WD algorithm employs 200 Gaussian probe

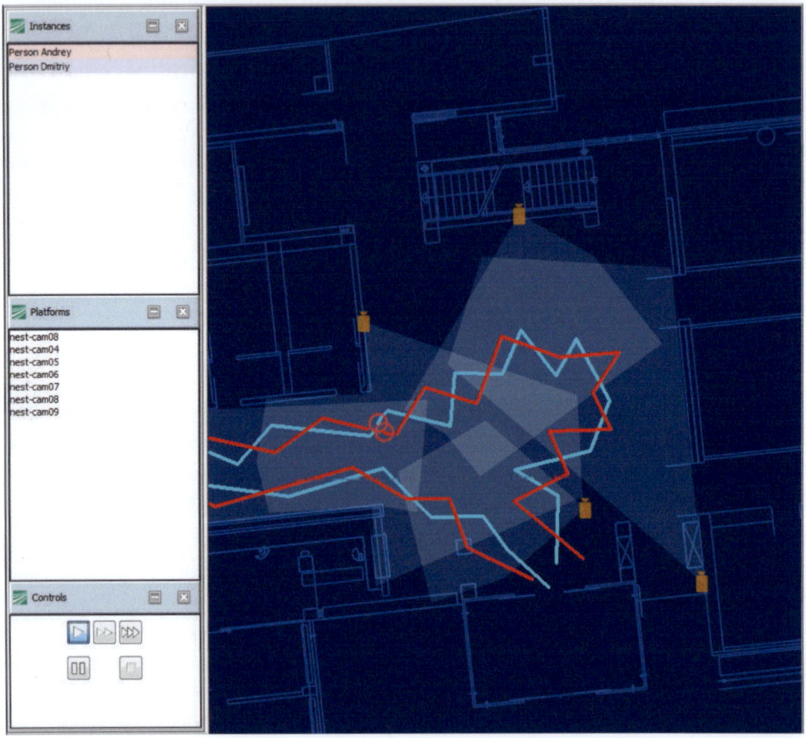


**Figure 4.14:** Tracking simulation: two persons (red circles) Andrey and Dmitriy are walking within a building. A set of cameras (yellow pictograms) are observing them with view coverages marked by lighter volumes. Andrey is wearing mostly red clothes and Dmitriy blue ones. The ground truth paths are presented by corresponding red and blue lines.

**(a)** No color consideration;

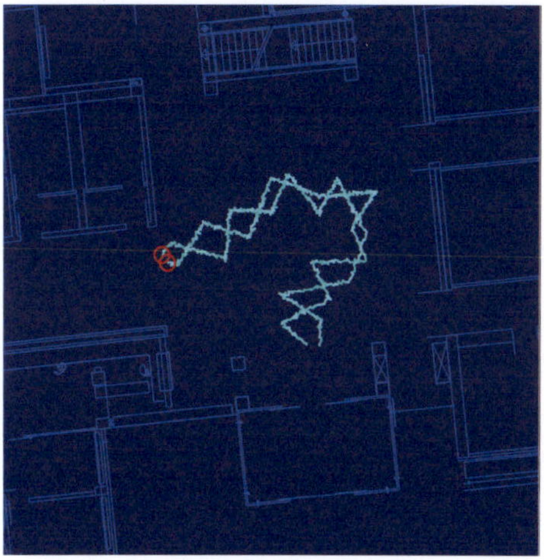

**(b)** Color attribute is considered;

**Figure 4.15:** Simulation scenario: recovered trajectories (particle filter results) without and with color estimation and matching.

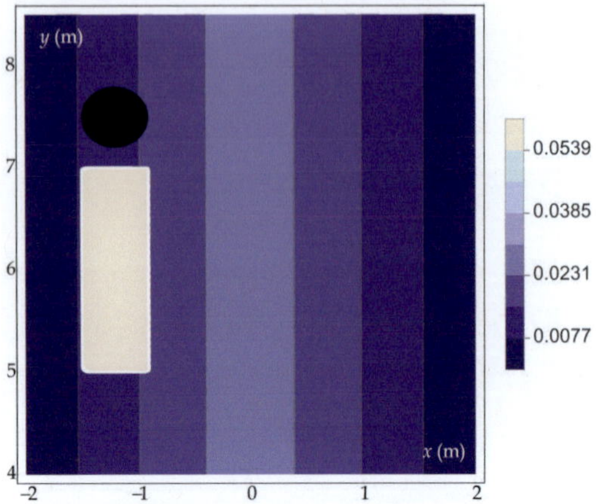


**(a)** Prior probability distribution of finding a person on an alley (view from above): black – a tree trunk, beige – a bench;

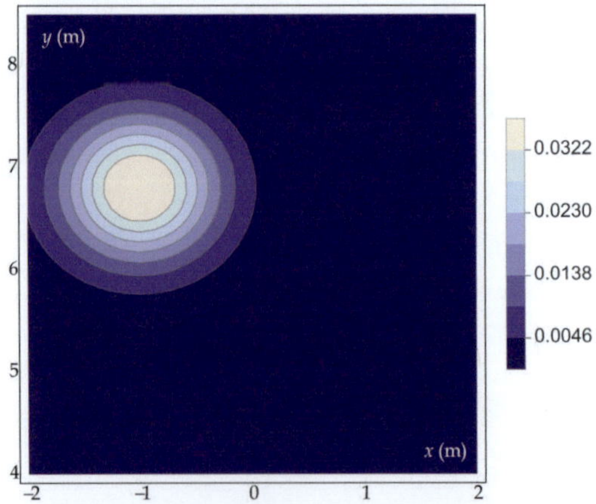


**(b)** Likelihood after an observation or state estimate of a person;

**Figure 4.16:** Bayesian fusion set-up: topology information and likelihood or state estimate.

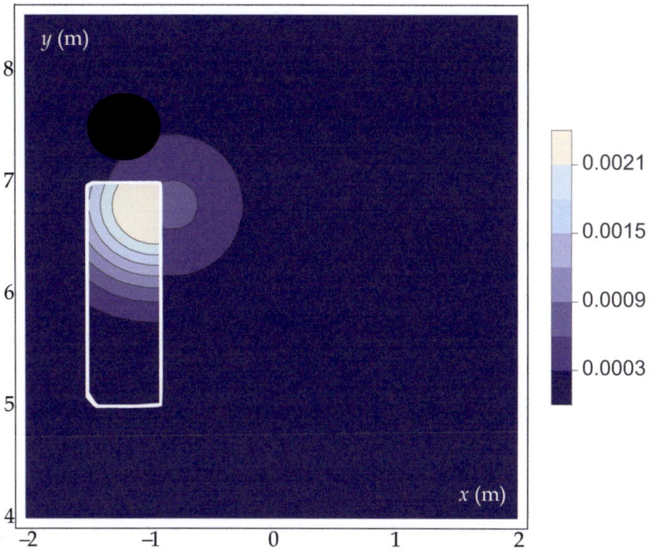


**Figure 4.17:** Bayesian fusion: a posterior state estimate benefits from the prior information, assigning more probability of finding a person to more plausible regions (e.g. in this case, the bench location).

functions with mean $m_i \propto N(\omega; W)$, blurring kernel size $E = 0.02$, window kernel position at $\omega = 1.0$, and size $W = 4.0$. The WC algorithm employs similar parameters.

The results of the reduction algorithms are presented in the Table 4.2. The symbols $\Sigma_{ISD}$ and $\Sigma_{WSD}$ denote symmetrized normalized ISD and WSD metrics [Hub09] defined as:

$$\Sigma_{\bullet} = \sqrt{\frac{d_{\bullet}(f,g)}{d_{\bullet}(f,0) + d_{\bullet}(0,g)}}, \tag{4.14}$$

where • stands for either ISD or WSD. The $d_{\text{ISD}}(f,g)$ is given in Eq. 2.22 and $d_{\text{WSD}}(f,g)$ in Eq. 2.38.

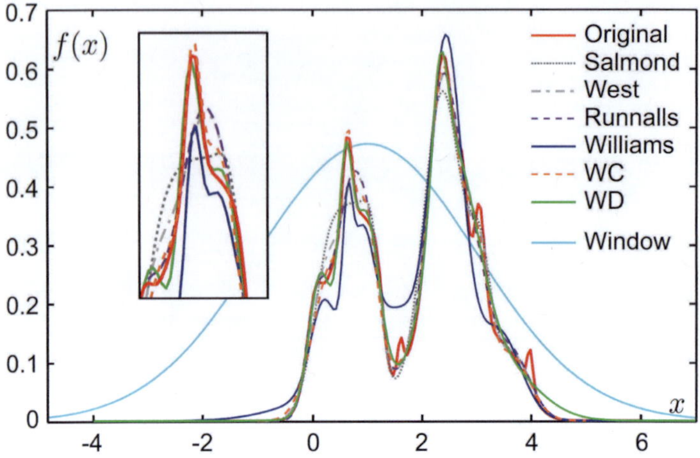


**Figure 4.18:** Comparison of GM reduction algorithms [Pak13].

## MHT with Weak Distance GM Reduction

In order to compare GM reduction algorithms over a prolonged time period, we have performed 2D multiple hypotheses tracking [Wil03b] with state estimate components reduction at each time iteration. The algorithms comparison is performed each time over the same data set. In order to have complicated cases in GM analysis, we have contaminated real observations with severe clutter: each observation is delivered with 100 false observations. Moreover, we demand at most 8 kernels to survive the reduction phase on each time step. In Fig. 4.19, we have plotted one test run of 50 time steps with WC reduction. For the overall comparison, we have performed 5 runs each of 50 steps for each reduction algorithm.

The WC and WD algorithms had the same run parameters: the measurement noise covariance matrix $\boldsymbol{C}_\upsilon = \mathrm{d}iag(1,1)$, the blurring kernel $\boldsymbol{E} = 49\boldsymbol{C}_\upsilon$, and the window function with mean $\vec{\boldsymbol{\omega}} = (200,200)^T$ and covariance matrix $\boldsymbol{W} = 4 \cdot 10^4 \boldsymbol{C}_\upsilon$. These parameters allow to limit the tasks with only 1000 probe

| Method | $\Sigma_{ISD}$ | $\Sigma_{WSD}$ | Runtime [s] | Mean | Variance |
|---|---|---|---|---|---|
| Original | - | - | - | 1.855 | 1.224 |
| Salmond | 0.0916 | 0.0518 | 3.90 | 1.855 | 1.224 |
| West | 0.0701 | 0.0295 | 0.017 | 1.855 | 1.224 |
| Runnalls | 0.0679 | 0.0271 | 0.090 | 1.855 | 1.224 |
| Williams | 0.1445 | 0.1239 | 2.477 | 1.838 | 1.197 |
| WC | 0.0547 | 0.0250 | 2.486 | 1.843 | 1.220 |
| WD | 0.0555 | 0.0102 | 62.47 | 1.886 | 1.302 |

**Table 4.2:** A comparison of GM reduction algorithms [Pak13]. The WD algorithm is the slowest approach (due to full-gradient search) but it greatly reduces the problem-adjusted WSD metric. The WC variant delivers small WSD and ISD. All square difference based algorithms change the mean and covariance during the reduction process. In the case of WD/WC it can be compensated by enforcing the equality of the first moments with corresponding probe functions with large weights. A larger WSD loss in WC algorithm compared to the original Williams' variant is not clear given the fact that our version employs Williams' minimization method based on WSD metric. We consider two possible reasons: imposed simulation conditions are too harsh, leading to GM degeneration with enormous discrepancy between weights of different components (a clue given by smaller $\Sigma_{WSD}$ average and per-step values of Runnalls' algorithm compared to Williams' algorithm). Second, the weak metric is possibly more sensitive to the sub-optimality of the Williams' initialization algorithm.

functions (due to approximate estimation of minimum sufficient number of probe functions around $\sqrt{\det \boldsymbol{E} / \det \boldsymbol{W}}$).

In spite of harsh conditions, almost all considered algorithms have managed to keep the track during the test runs. The only algorithm that had lost the track during two runs was the Pruning algorithm, which represents simple removal of kernels with the smallest weights. However, the number of kernels in the estimate distribution just before each reduction step was significantly different between algorithms, which can be considered for further algorithms improvements. Moreover, the approximation quality had varied from algorithm to algorithm. The results are presented in Table 4.3 in a logarithmic scale.

### 4.4.4 Bestimmtheit Assessment

In [Bel12b] we discuss Bestimmtheit (refer to Sec. 3.3.3) assessment for a kitchen environment. Let us move now to a living room and consider a task of watering a recently brought small ficus on a chest of drawers (Figure 4.20a). The experimental parameters are listed in the Table 4.4. For simplicity we consider only marginal probability distributions. The chest of drawers is a persistent entity of constant size (0.80; 1.00; 0.48) m and constant position, so, it is defined in the prior knowledge. Let us suppose that its representative in the dynamic model is already correctly classified. This imposes prior restrictions on the geometrical attributes of the flowerpot: it has to be on the top of the drawer. The flower entity is new to the ARMAR-III robot (brought a few minutes ago). The robot, which performs the watering task, has no explicit prior information about possible flowerpot sizes. The placeholder surface limits the prior size of the flowerpot to (0.01-0.80; 0.01-1.50; 0.01-0.48) m. Here we assume the size at least of 1 cm and at most of the surface area (chest of drawers' surface and the height of maximum 1.5 m). Since the exact prior size is unknown, the prior position is allowed to be (0.00-0.80; 0.00; 0.00-0.48) m with the reference point of the right farthest corner of

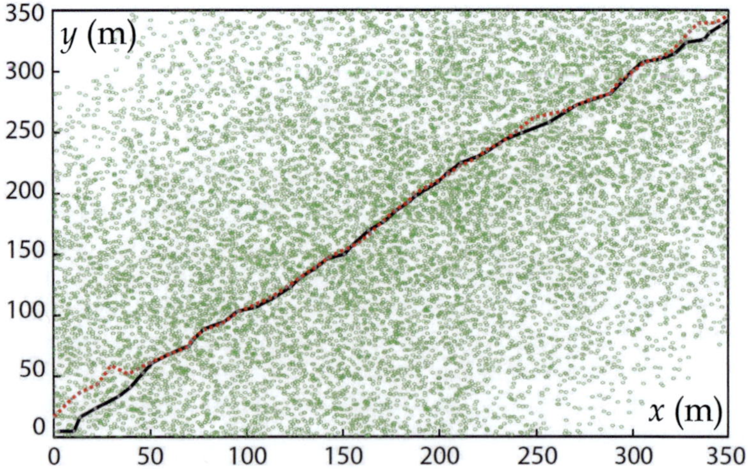


**Figure 4.19:** 2D tracking [Pak13]: ground truth (black), real and clutter observations (green), WC algorithm estimate (red).

| Method | $\langle \log_{10}(\Sigma_{WSD}) \rangle$ | $\langle \log_{10}(\Sigma_{ISD}) \rangle$ |
|---|---|---|
| Prune | -2.4979 | -2.1405 |
| West | -2.9392 | -1.4951 |
| Salmond | -3.2378 | -1.2886 |
| Runnalls | -4.0232 | -3.3137 |
| Williams | -3.9465 | -3.5596 |
| WC | -1.8776 | -1.6956 |
| WD | -3.7910 | -3.1089 |

**Table 4.3:** A comparison of common GM reduction methods in 2D MHT tracking [Pak13]: the averaged logarithms of the WSD and the ISD values. It is surprising to get worse weak squared difference (WSD) value in WC algorithm compared to William's algorithm, since we had used the William's algorithm for minimization of the WSD metric. The reason for this is in almost degenerated GMs with huge discrepancies between kernel weights arising from the harsh tracking conditions. Moreover, the weak metric algorithm is very sensitive to the William's initialization method. The WD algorithm seems to be at the level of the state of the art algorithms with a potential for improvements (due to problem- and sensor-specific parametrization and ability to handle generalized functions). It is important, though, to find those application fields, as well as optimal sets of probe function, which will benefit most from the proposed approach.

the placeholder. For simplicity, we assume flat p.d.f.s within allowed values (though, as discussed previously, we normally use Gaussian distributions).

The posterior distribution can be obtained with Bayesian fusion of observations with prior state. Let us suppose a Gaussian form for the probability distributions. The ARMAR robot has stereo cameras and, thus, can perform a 3D spacing of the presented scene, though with a large uncertainty. A typical spatial uncertainty (standard deviations for marginal spatial distributions) for a flowerpot placed on the 1 m height and 1 m away in front of the robot is around (0.05; 0.06; 0.09) m depending on calibration and sight of view angles. Since the task involves pouring of water over the furniture, we require relatively fine granularity of modeling information. In this case, the least discernible quantum (LDQ) is limited mostly by sensors resolution uncertainty.

The prior and posterior $x$ and $z$ distributions are presented in Figures 4.20b and 4.20c. The distribution for $y$ is assumed to be a delta-function (flowerpot stands exactly on the chest of drawers' surface). Similarly, the sizes of the flowerpot bounding box are Gaussian p.d.f.s with parameters listed in Table 4.4. The resulting information entropies, KL distances between prior and posterior and Bestimmtheit are presented in Table 4.5.

| Chest of drawers | |
|---|---|
| size $d_x$ | 0.80 |
| size $d_y$ | 1.00 |
| size $d_z$ | 0.48 |
| **Flowerpot** | |
| prior position $x$ | $\in [0.00; 0.80]$ |
| prior position $y$ | 0.00 |
| prior position $z$ | $\in [0.00; 0.48]$ |
| prior size $d_x$ | $\in [0.01; 0.80]$ |
| prior size $d_y$ | $\in [0.01; 1.50]$ |
| prior size $d_z$ | $\in [0.01; 0.48]$ |
| posterior position $x$ | $N(0.02, 0.05)$ |
| posterior position $y$ | 0.00 |
| posterior position $z$ | $N(0.05, 0.09)$ |
| posterior size $d_x$ | $N(0.18, 0.05)$ |
| posterior size $d_y$ | $N(0.10, 0.06)$ |
| posterior size $d_z$ | $N(0.08, 0.09)$ |
| **Camera parameters** | |
| $\sigma_x$ | 0.05 |
| $\sigma_y$ | 0.06 |
| $\sigma_z$ | 0.09 |
| **LDQ** | |
| LDQ $x,y,z$ | equal to camera $\sigma$ |

**Table 4.4:** Example parameters for Bestimmtheit assessment.

In the common case (no given tasks $\Rightarrow$ equal weight coefficients), the resulting Bestimmtheit over geometry attribute descriptions is 0.61 with position $z$ and sizes $x$ and $z$ marked for further exploration. We interpret this Bestimmtheit value as geometry information observed to 61% of potentially measurable/relevant information. This value, however, depends on the LDQ (and thus, on actual visual information uncertainty). For example, if the

| Attribute | Entropy | | KL distance | Bestimmtheit |
|---|---|---|---|---|
| | prior | posterior | | |
| position $x$ | 2.77 | 0.88 | 0.94 | 0.68 |
| position $y$ | 0.00 | 0.00 | 0.00 | 1.00 |
| position $z$ | 1.74 | 0.93 | 0.26 | 0.46 |
| size $x$ | 2.77 | 1.46 | 1.30 | 0.48 |
| size $y$ | 3.4 | 1.23 | 1.89 | 0.64 |
| size $z$ | 2.28 | 1.33 | 0.83 | 0.42 |

**Table 4.5:** Entropy, KL distance between prior and posterior and Bestimmtheit assessment.

robot would stand closer to the chest of drawers so that uncertainty would be 20% less, the Bestimmtheit would be 0.59.

Let us consider the sufficiency of Bestimmtheit value for a given task, namely, for watering the ficus. For simplicity, we calculate a limit for the $z$-position. To this end, we put three sigma precision to be of 2.5 cm in order to stay within the 5 cm area around the plant to the border of the flowerpot. Hereby, the position for watering in a discretized distribution form has to be within the 5 cm area. In the "worst acceptable" case, the position distribution is flat over this area, giving a maximum allowed entropy $H^*$. In this case, we require the Bestimmtheit value of $B = 1 - H^* / H_{\max} = 1 - 1.18/3.91 = 0.70$. Hence, the Bestimmtheit is still insufficient for the required task and perception subsystem is triggered for further flowerpot observation.

## 4.4.5 Prior Knowledge Matching

*1D Simulation* We have compared two state of the art distances: KL and Wasserstein $L^2$ (refer to Sec. 3.5.2) within a 1D simulator. The test set-up was implied to be a kitchen environment with plausible entity sizes and camera parameters. Hereby, the prior knowledge had contained $K$ concepts, each presented by a type label and a size distribution $N(\mu_c; \sigma_c)$, where $\mu_c \in [0.05; 3.00]$ m and $\sigma_c \in [0.05; 0.20]$ m. The dynamic model was populated with $R = K \cdot F$ representatives of the same structure: type label from

**(a)** Flowerpot;

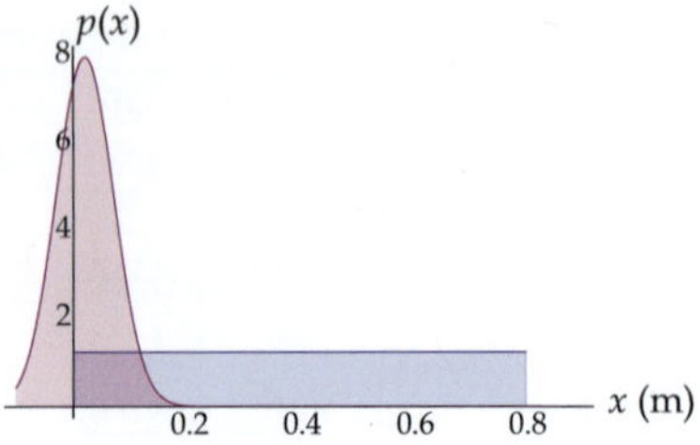


**(b)** Position $x$;

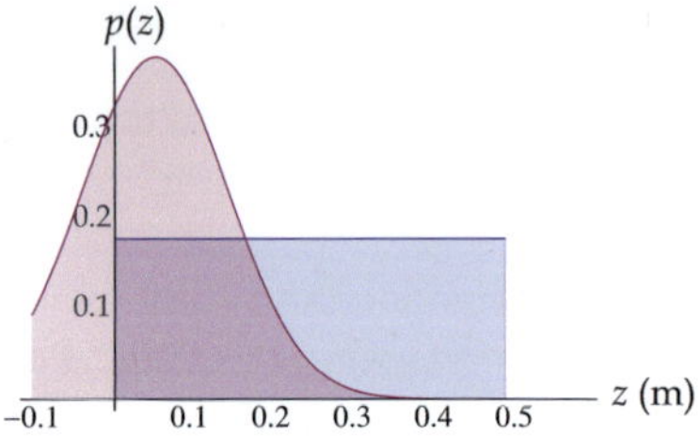


**(c)** Position $z$;

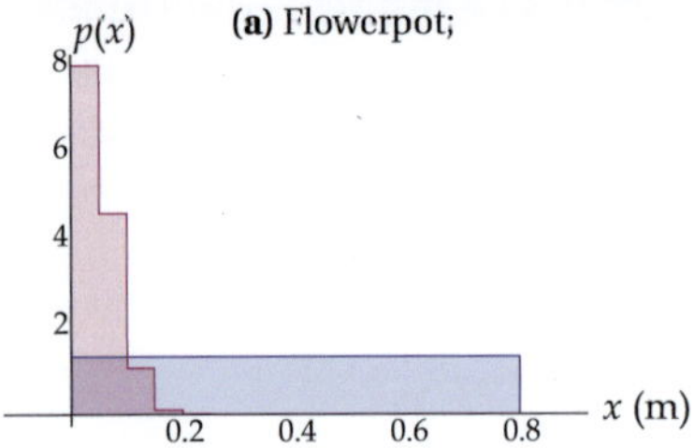


**(d)** Discretization of position $x$;

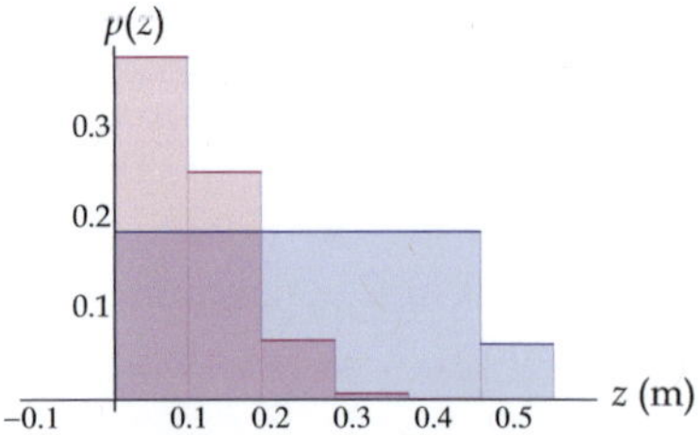


**(e)** Discretization of position $z$;

**Figure 4.20:** Experiment data: (a) – flowerpot for watering task, (b)-(c) – $x$ and $z$ positions of the flowerpot relative to the chest of drawers corner, (d)-(e) – LDQ discretization. Prior probability density functions are presented in blue and posterior in pink color.

prior knowledge and size distribution $N(\mu_r;\sigma_r)$, where $\mu_r$ was chosen randomly within the area of $1\sigma_c$ and $\sigma_r$ within [0.005; 0.010] m. In this set-up, $F$ representatives were correctly pre-matched to each original, creating all conditions for distances comparison. In Fig. 4.21 we have depicted an example of random prior knowledge distributions and corresponding likelihoods generated according to prior parameters.

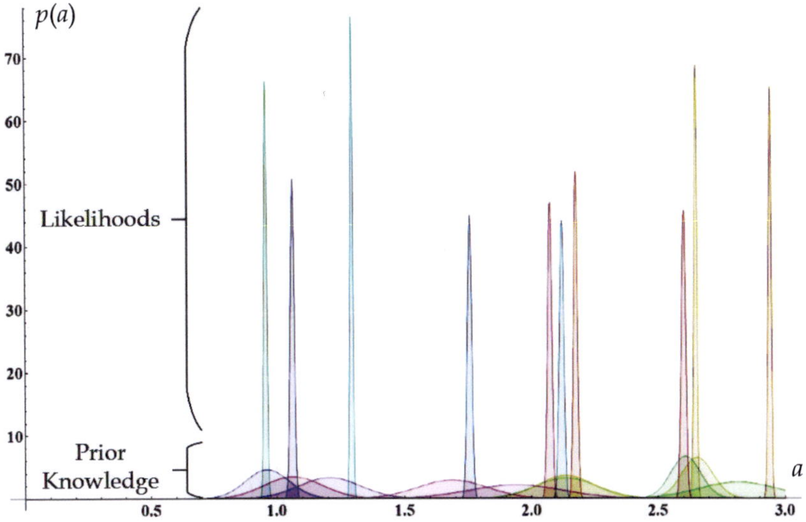


**Figure 4.21:** Simulation data for 10 pairs of likelihoods and corresponding prior knowledge distributions.

For better performance of representative-to-concept matching, it is advantageous to employ closed forms of the distances. For example, the KL divergence contains $p(a)\log\left(p(a)/q(a)\right)\, da$ that leads to narrow delta-function-similar functions in the integral, not easily to compute numerically. The closed forms for KL divergence and Wasserstein $L^2$ are found as:

$$d_{KL}(p,q) = \int\limits_{-\infty}^{+\infty} p(x)\log\frac{p(x)}{q(x)}dx$$
$$= \int\limits_{-\infty}^{+\infty} \frac{\exp\left(-\frac{(x-\mu_p)^2}{2\sigma_p^2}\right)\left(-\frac{(x-\mu_p)^2}{2\sigma_p^2}+\frac{(x-\mu_q)^2}{2\sigma_q^2}-\log\sigma_p+log\sigma_q\right)}{\sqrt{2\pi}\sigma_p}$$
$$= \frac{(\mu_p-\mu_q)^2+(\sigma_p-\sigma_q)(\sigma_p+\sigma_q)+2\sigma_q^2(-\log\sigma_p+\log\sigma_q)}{2\sigma_q^2}; \tag{4.15}$$

$$d_{L^2}(p_1,p_2) = \int\left((\acute{p}_1(x)-\acute{p}_2(x)\right)^2 dx$$
$$= \left(\frac{\exp\left(-\frac{(x-\mu_{p_1})^2}{2\sigma_{p_1}^2}\right)}{\pi^{1/4}\sqrt{\sigma_{p_1}}} - \frac{\exp\left(-\frac{(x-\mu_{p_2})^2}{2\sigma_{p_2}^2}\right)}{\pi^{1/4}\sqrt{\sigma_{p_2}}}\right)^2, \tag{4.16}$$
$$\text{with } \acute{p}_i = \frac{p_i(x)}{\sqrt{\int p_i(x)^2 dx}}.$$

The resulting plots for distances comparison are given in Fig. 4.22-4.24 with different parameters $K$ and $R$. A difference of correct matching cases of KL minus correct matching of $L^2$ ranges within 2.5%.

### An Example of Real Matching

*Set-Up* In [Bel12c] we consider a scenario of an exploration of entities on a table during a coffee break. Let us extend it to a breakfast episode. The table is a persistent entity, thus, its size and location is pre-defined in the prior knowledge. A typical spatial uncertainty coming from the ARMAR cameras for a kettle placed 1.5 m in front of the robot is about 0.1 m depending on the angular calibration and the angle of sight [Bel12b]. For simplification,

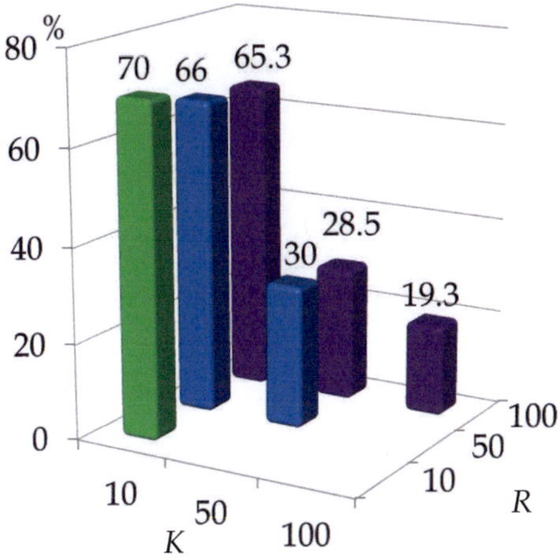


**(a)** Correct matching with KL;

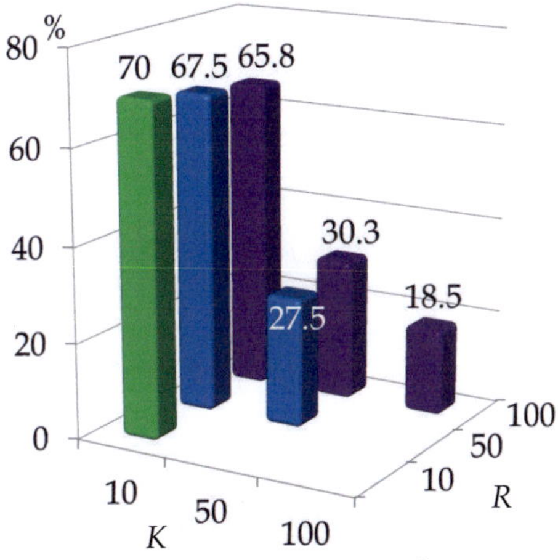


**(b)** Correct matching with $L^2$;

**Figure 4.22:** Simulation results of distances comparison: correct matching. $R$ – number of representatives in a dynamic model, $K$ – number of concepts in the prior knowledge.

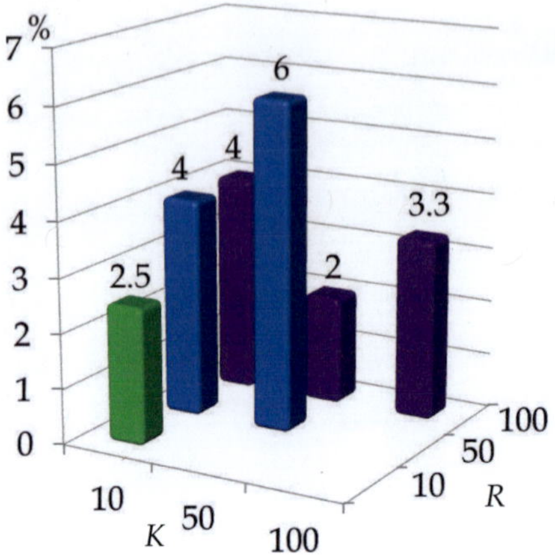


**(a)** Correct matching with KL when $L^2$ is wrong;

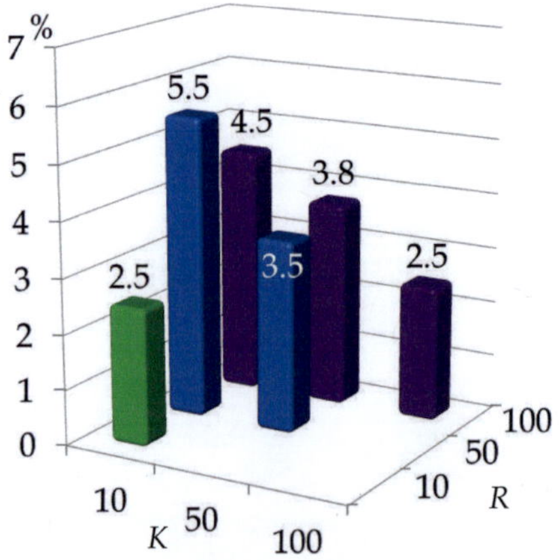


**(b)** Correct matching with $L^2$ when KL is wrong;

**Figure 4.23:** Simulation results of distances comparison: correct matching when other metric makes error. $R$ – number of representatives in a dynamic model, $K$ – number of concepts in the prior knowledge.

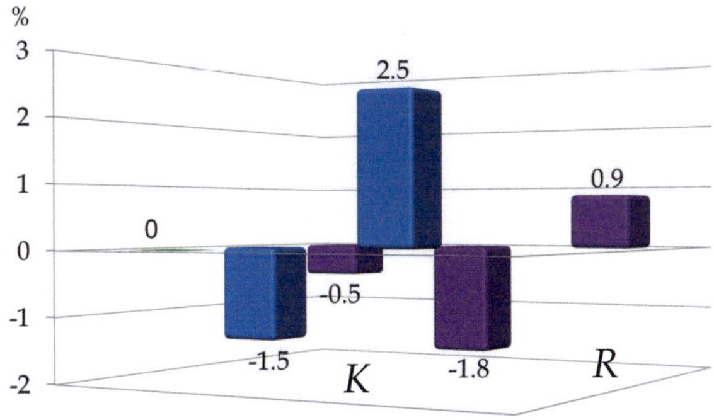


**Figure 4.24:** Simulation results of distances comparison: correct matching KL-$L^2$. $R$ – number of representatives in a dynamic model, $K$ – number of concepts in the prior knowledge.

we allow only a limited set of entities: specific kettle, tea cups and a cucumber. The choice of the entities is special: the kettle is a known entity with recognizable noise production upon boiling, cups are represented by a tea cup concept without prior statistics and a cucumber is represented by a class concept with color description. Moreover, cucumber, cup and kettle have the same shape CYLINDER; the cup and the kettle both have a handle. An example of an experimental set-up is presented in Figure 4.25. As mentioned in Section 4.4.4, typical standard deviations for marginal spacial distributions in such a set-up are about (0.05; 0.06; 0.09) m. Hereby, we can immediately meet several problems:

1. entities merge – entities placed too close to each other (less than 10 cm distance) cannot be observed correctly;

2. non-isotropic lighting – emerging shadows are segmented as a part of an entity;

3. color balance – colors are not constant within spatio-temporal sections due to non-isotropic lighting (a cucumber is almost black on the left side and green on the right side) and changing lighting (e.g. daylight from windows and artificial lighting during the nighttime).

**Figure 4.25:** Scene example for matching.

Since we focus here on the proposed matching approach, we solve these issues as following: the entities merge is avoided by placing entities far enough to each other. The latter two problems are solved by constant artificial lighting directed from all sides and curtained windows. Moreover, we simplify the scene by neglecting bounding box sizes for cylinder-like entities (e.g. cups) and consider cylinders (i.e. radius and height of the entities). Finally, due to practical considerations, we represent color attribute description in form of normalized color histogram (number of pixels with RGB colors corresponding to each bin divided by the total number of pixels).

*Dynamic Model and Prior Knowledge Information* For this experimental set-up we have 4 representatives (REPR_1, REPR_2, REPR_3, REPR_4) corresponding to 4 entities (BREAD, CUCUMBER, TEA CUP, AEG KETTLE) and 3 concepts in prior knowledge (CONCEPT_CLASS_CUCUMBER, CONCEPT_CLASS_TEA_CUP, CONCEPT_ENTITY_AEG_KETTLE). The attribute descriptions of representatives and corresponding concepts are presented in Tables 4.6-4.9.

| Entity BREAD | | |
|---|---|---|
| | REPR_1 | no match in prior knowledge |
| SHAPE | 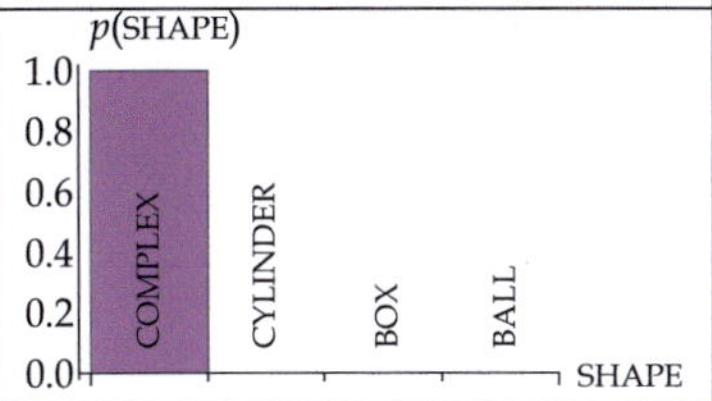 | - |
| COLOR | 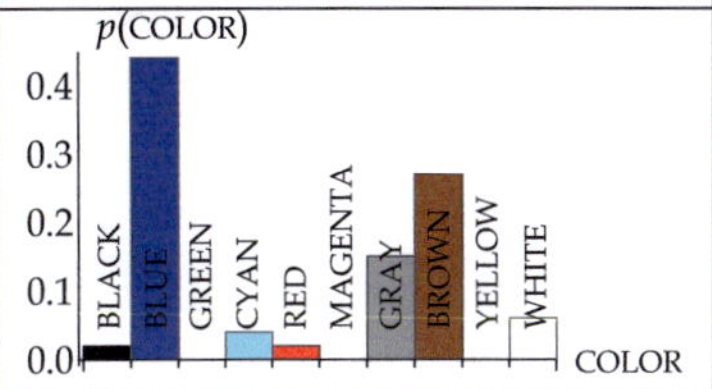 | - |
| SIZE_DX | 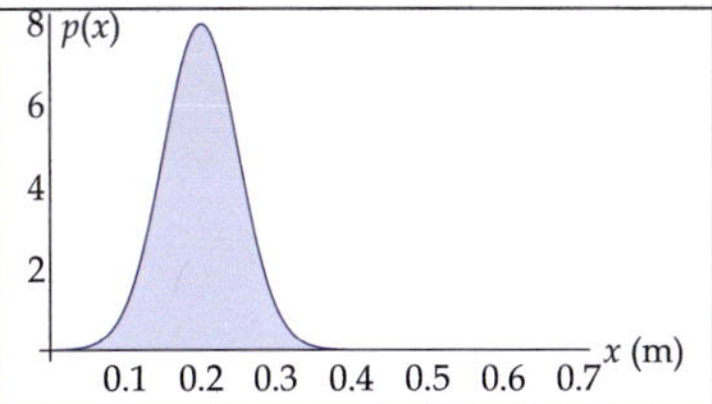 | - |
| SIZE_DY | 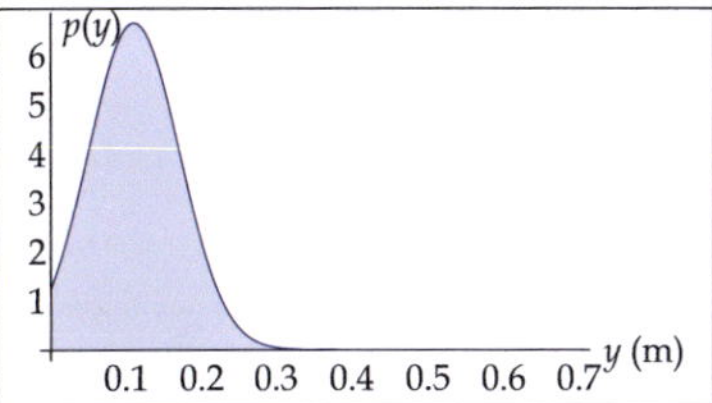 | - |
| SIZE_DZ | 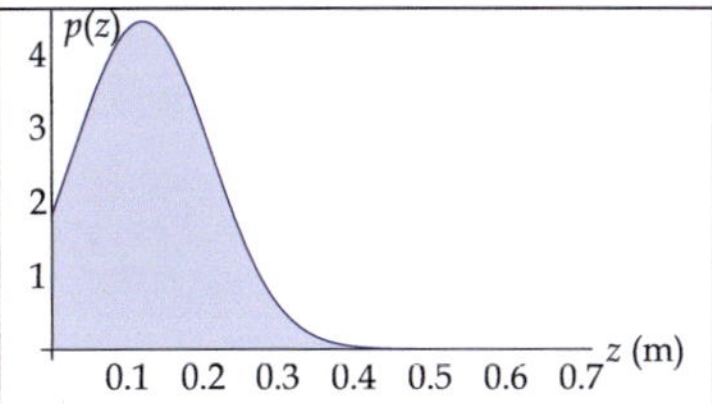 | - |

**Table 4.6:** Real matching example: entity BREAD attribute descriptions.

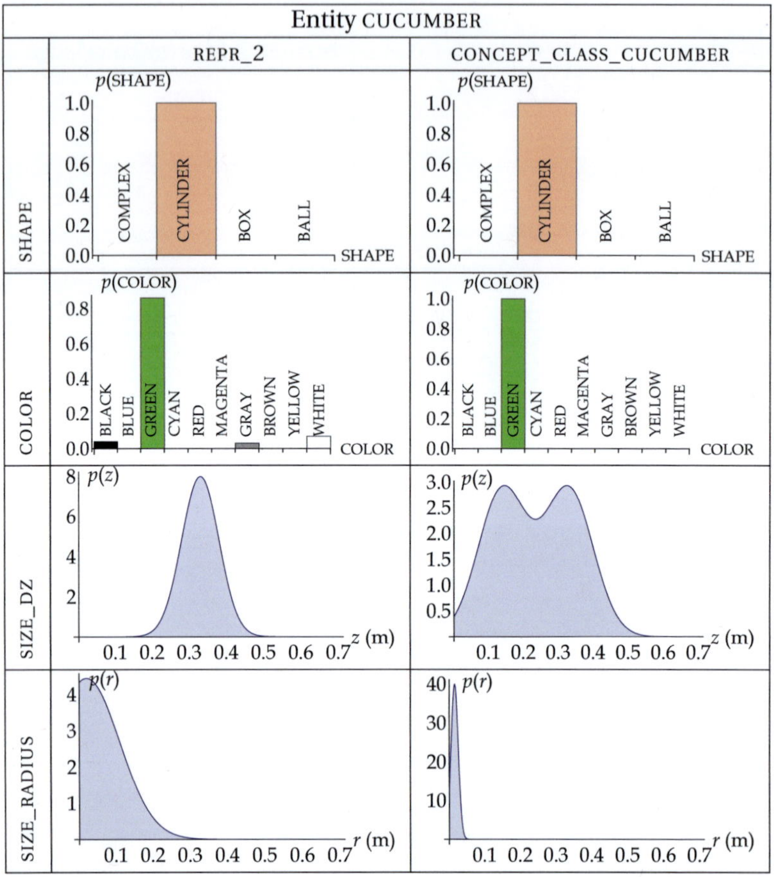


**Table 4.7:** Real matching example: entity CUCUMBER attribute descriptions and prior knowledge. Length SIZE_DZ distribution describes two sorts of cucumbers.

| Entity TEA CUP | | |
|---|---|---|
| | REPR_3 | CONCEPT_CLASS_TEA_CUP |
| SHAPE | 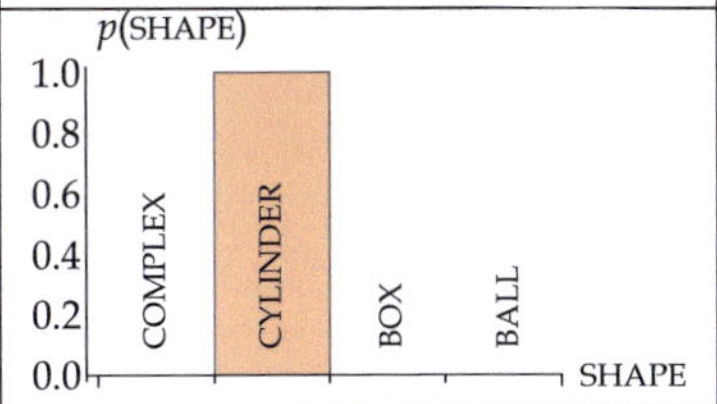 | 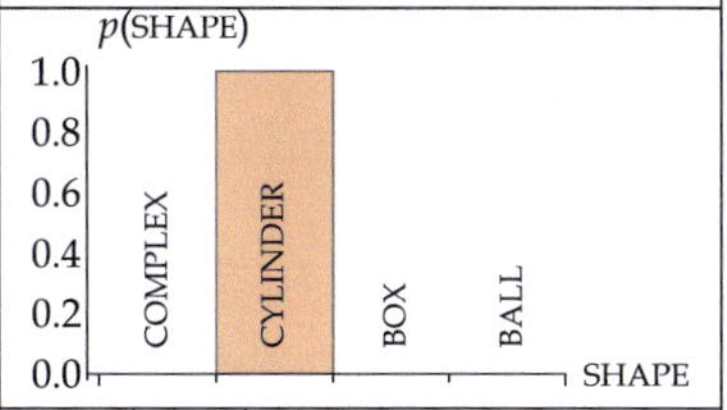 |
| COLOR | 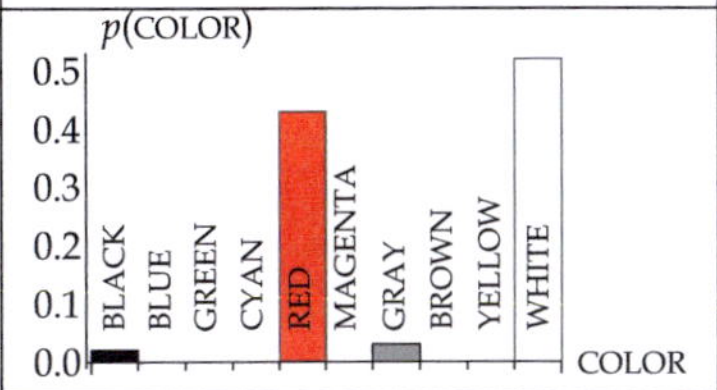 | 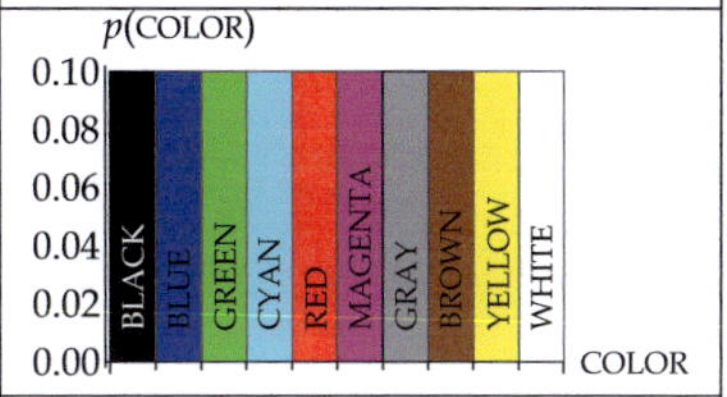 |
| SIZE_DZ | 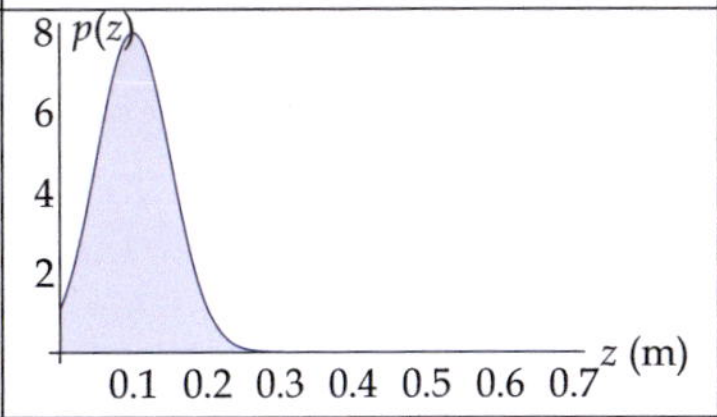 | 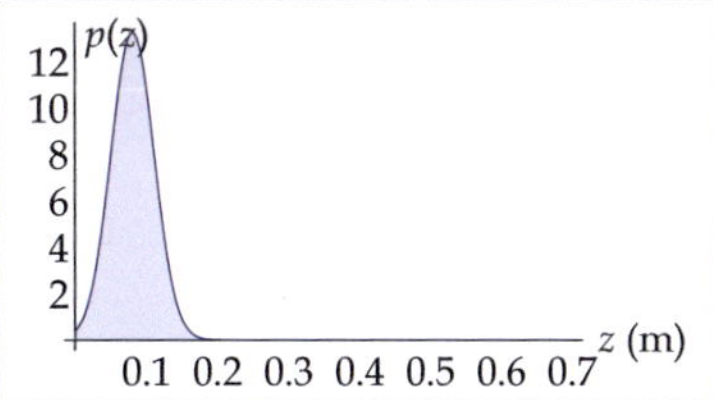 |
| SIZE_RADIUS | 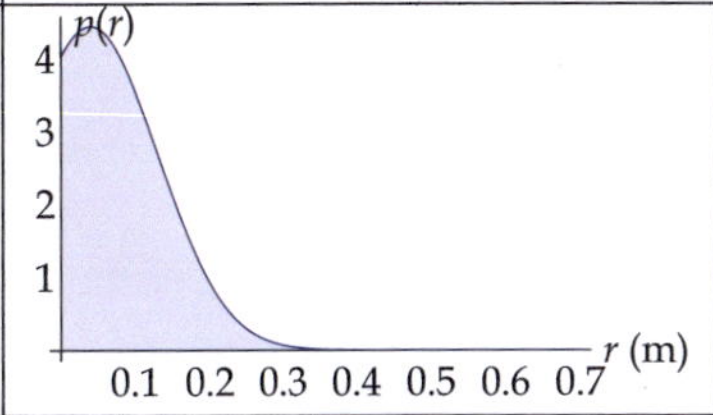 | 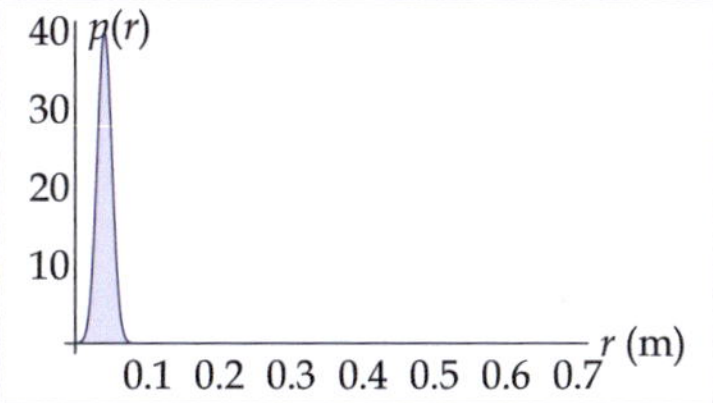 |
| HAS_HANDLE | 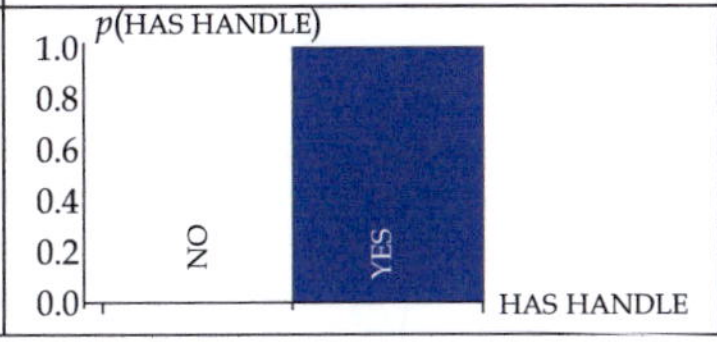 | 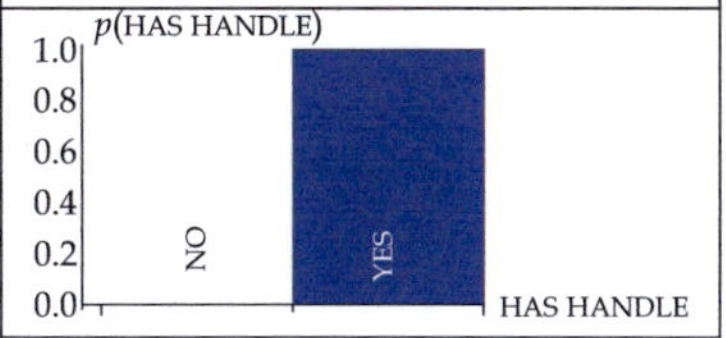 |

**Table 4.8:** Real matching example: entity TEA CUP attribute descriptions and prior knowledge.

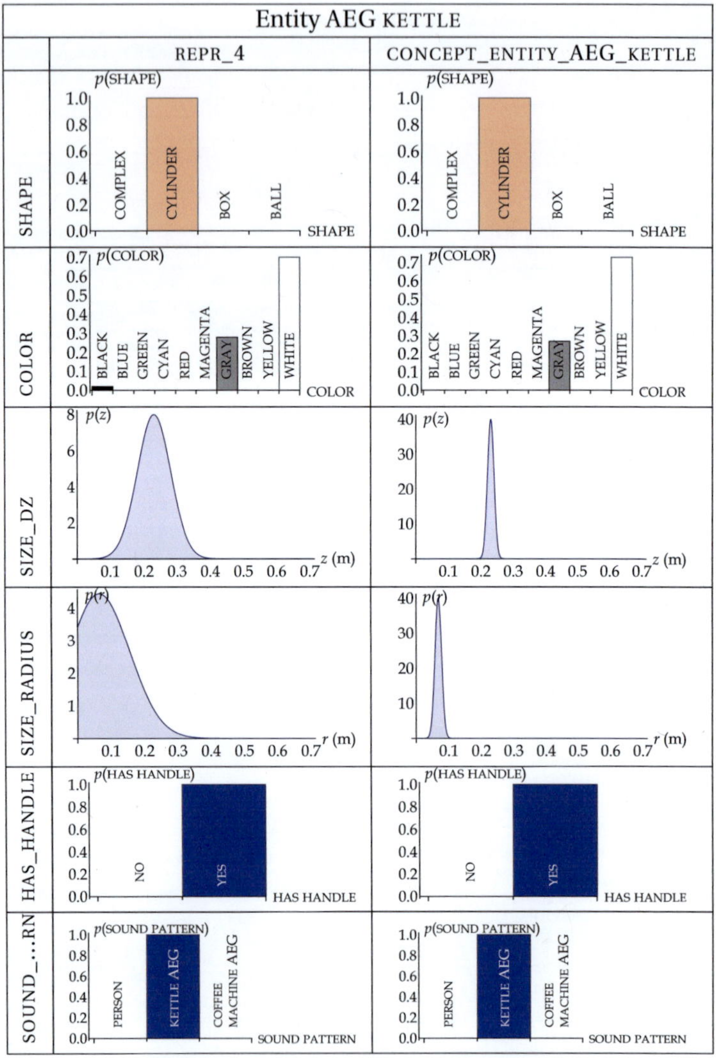


**Table 4.9:** Real matching example: entity AEG KETTLE attribute descriptions and prior knowledge.

*Matching Example* The structural difference analysis results are presented in Table 4.10. The value difference analysis results are presented in Table 4.11.

| | REPR_1 | REPR_2 | REPR_3 | REPR_4 |
|---|---|---|---|---|
| CONCEPT...CUCUMBER | 0.50 (0.33) | 0 (0) | 0.2 (0.11) | 0.33 (0.2) |
| CONCEPT...TEA_CUP | 0.57 (0.4) | 0.2 (0.11) | 0 (0) | 0.17 (0.09) |
| CONCEPT...KETTLE | 0.63 (0.45) | 0.33 (0.2) | 0.17 (0.09) | 0 (0) |

**Table 4.10:** Structural difference analysis results: Tanimoto-metric distance (numbers) and inverse Sørenson index distance (numbers in braces).

We can see that the dynamic model contains systematic errors (sensor error, color reflections, etc.) that can dramatically affect distribution matching. For example, due to lighting peculiarities (leading to shadows and flares) and other effects, the representative REPR_2 (entity CUCUMBER) has black, gray and white components in the color histogram, while corresponding concept has only the green component. If we compute KL divergence from this couple of distributions, the result will be infinite, since prior distribution contains zeroes at black, gray and white bins. This infinity drastically affects the overall distance between REPR_2 and CONCEPT_CLASS_CUCUMBER, leading to minimal similarity. At this point, the cucumber model has to be adjusted in such way that all other colors are allowed with a small probability. For example, we can introduce a flat distribution component with very low probability value. Alternatively, we can ignore an amount of values in representative's distributions equal to empirically found values of systematic effects (e.g. employ a 7 per cent threshold).

Relatively large values and discrepancies in the Table 4.11 come from poor sensors' performance. In the case of large uncertainty, it can be impossible to match to concepts of lower ontology levels. The sufficiency of information, as well as lower matching bound can be determined by Bestimmtheit calculation. If matching is not possible, we stop at a parent concept (e.g. CYLINDER instead of TEA CUP) during the depth search.

Now we normalize calculated distances. For simplicity, let us put normalizing functional as follows:

$$f_{DoB} := 1 - \frac{1}{2^{|d_{DoB}(A)|}}. \tag{4.17}$$

| SHAPE | | | | |
|---|---|---|---|---|
| | REPR_1 | REPR_2 | REPR_3 | REPR_4 |
| CONC...CUCUMBER | ∞ (2) | 0 (0) | 0 (0) | 0 (0) |
| CONC...TEA_CUP | ∞ (2) | 0 (0) | 0 (0) | 0 (0) |
| CONC...KETTLE | ∞ (2) | 0 (0) | 0 (0) | 0 (0) |
| COLOR | | | | |
| | REPR_1 | REPR_2 | REPR_3 | REPR_4 |
| CONC...CUCUMBER | ∞ (0.34) | 0.13 (0.01) | ∞ (0.48) | ∞ (0.58) |
| CONC...TEA_CUP | 0.98 (0.18) | 1.85 (0.69) | 1.49 (0.3) | 1.65 (0.4) |
| CONC...KETTLE | ∞ (0.32) | ∞ (0.86) | ∞ (0.23) | -0.02 (0) |
| SIZE_DX | | | | |
| CONC...CUCUMBER | - (-) | - (-) | - (-) | - (-) |
| CONC...TEA_CUP | - (-) | - (-) | - (-) | - (-) |
| CONC...KETTLE | - (-) | - (-) | - (-) | - (-) |
| SIZE_DY | | | | |
| CONC...CUCUMBER | - (-) | - (-) | - (-) | - (-) |
| CONC...TEA_CUP | - (-) | - (-) | - (-) | - (-) |
| CONC...KETTLE | - (-) | - (-) | - (-) | - (-) |
| SIZE_DZ | | | | |
| CONC...CUCUMBER | 0.56 (0.47) | 0.68 (0.61) | 0.92 (0.82) | 0.66 (0.61) |
| CONC...TEA_CUP | 3.79 (0.56) | 35.1 (2) | 0.60 (0.22) | 12.88 (1.93) |
| CONC...KETTLE | 98.3 (1.52) | 60.39 (1.82) | 94.89 (1.95) | 10.39 (0.76) |
| SIZE_RADIUS | | | | |
| CONC...CUCUMBER | - (-) | 37.93 (0.73) | 40.93 (0.87) | 50.3 (1.07) |
| CONC...TEA_CUP | - (-) | 39.8 (0.71) | 37.8 (0.8) | 40.93 (0.95) |
| CONC...KETTLE | - (-) | 48.85 (0.8) | 41.45 (0.84) | 37.82 (0.91) |
| HAS_HANDLE | | | | |
| CONC...CUCUMBER | - (-) | - (-) | - (-) | - (-) |
| CONC...TEA_CUP | - (-) | - (-) | 0 (0) | 0 (0) |
| CONC...KETTLE | - (-) | - (-) | 0 (0) | 0 (0) |
| SOUND_PATTERN | | | | |
| CONC...CUCUMBER | - (-) | - (-) | - (-) | - (-) |
| CONC...TEA_CUP | - (-) | - (-) | - (-) | - (-) |
| CONC...KETTLE | - (-) | - (-) | - (-) | 0 (0) |

**Table 4.11:** Value difference analysis results: KL divergence (numbers) and Wasserstein $L^2$ distance (numbers in braces). Missing parameters (due to restrictions of the attributes scheme or missing corresponding attributes) are denoted by the "-"-sign.

The normalized distances are presented in Table 4.12.

| SHAPE | | | | |
|---|---|---|---|---|
| | REPR_1 | REPR_2 | REPR_3 | REPR_4 |
| CONCEPT...CUCUMBER | 1 (0.75) | 0 (0) | 0 (0) | 0 (0) |
| CONCEPT...TEA_CUP | 1 (0.75) | 0 (0) | 0 (0) | 0 (0) |
| CONCEPT...KETTLE | 1 (0.75) | 0 (0) | 0 (0) | 0 (0) |
| COLOR | | | | |
| | REPR_1 | REPR_2 | REPR_3 | REPR_4 |
| CONCEPT...CUCUMBER | 1 (0.21) | 0.09 (0.01) | 1 (0.28) | 1 (0.33) |
| CONCEPT...TEA_CUP | 0.49 (0.12) | 0.72 (0.38) | 0.64 (0.19) | 0.36 (0.24) |
| CONCEPT...KETTLE | 1 (0.2) | 1 (0.45) | 1 (0.15) | 0.01 (0) |
| SIZE_DZ | | | | |
| CONCEPT...CUCUMBER | 0.32 (0.11) | 0.38 (0.35) | 0.47 (0.44) | 0.37 (0.35) |
| CONCEPT...TEA_CUP | 0.93 (0.32) | 1 (0.75) | 0.34 (0.14) | 1 (0.74) |
| CONCEPT...KETTLE | 1 (0.65) | 1 (0.72) | 1 (0.74) | 1 (0.41) |
| SIZE_RADIUS | | | | |
| CONCEPT...CUCUMBER | - (-) | 1 (0.4) | 1 (0.45) | 1 (0.52) |
| CONCEPT...TEA_CUP | - (-) | 1 (0.39) | 1 (0.43) | 1 (0.48) |
| CONCEPT...KETTLE | - (-) | 1 (0.43) | 1 (0.44) | 1 (0.47) |
| HAS_HANDLE | | | | |
| CONCEPT...CUCUMBER | - (-) | - (-) | - (-) | - (-) |
| CONCEPT...TEA_CUP | - (-) | - (-) | 0 (0) | 0 (0) |
| CONCEPT...KETTLE | - (-) | - (-) | 0 (0) | 0 (0) |
| SOUND_PATTERN | | | | |
| CONCEPT...CUCUMBER | - (-) | - (-) | - (-) | - (-) |
| CONCEPT...TEA_CUP | - (-) | - (-) | - (-) | - (-) |
| CONCEPT...KETTLE | - (-) | - (-) | - (-) | 0 (0) |

**Table 4.12:** Normalized value difference analysis results: KL divergence (numbers) and Wasserstein $L^2$ distance (numbers in braces). Missing parameters (due to restrictions of the attributes scheme or missing corresponding attributes) are denoted by the "-"-sign.

We sum the resulting distances up and normalize the sum with $f_v := 1/N_{\{A\}_r \cap \{A\}_c}$. The resulting value differences are presented in Table 4.13.

Let us assign $\lambda_s = \lambda_v = 0.5$. In this case, the final distances between representatives and concepts is presented in Table 4.14. As we can see, Tanimoto-KL and Sørenson-$L^2$ approaches give equivalent results.

The Maximum A Posteriori (MAP) classification applied to Tanimoto-KL method with classification threshold of 0.25 gives us the following matching scheme:

- REPR_1 → ? (correct);
- REPR_2 → CONCEPT_CLASS_CUCUMBER (correct);
- REPR_3 → CONCEPT_CLASS_TEA_CUP (correct);
- REPR_4 → CONCEPT_ENTITY_AEG_KETTLE (correct).

| | REPR_1 | REPR_2 | REPR_3 | REPR_4 |
|---|---|---|---|---|
| CONCEPT...CUCUMBER | 0.77 (0.36) | 0.37 (0.19) | 0.62 (0.29) | 0.59 (0.3) |
| CONCEPT...TEA_CUP | 0.81 (0.4) | 0.68 (0.38) | 0.4 (0.15) | 0.47 (0.29) |
| CONCEPT...KETTLE | 1 (0.53) | 0.75 (0.4) | 0.6 (0.27) | 0.34 (0.15) |

**Table 4.13:** Normalized value difference $d_v$: KL divergence (numbers) and Wasserstein $L^2$ distance (numbers in braces).

| | REPR_1 | REPR_2 | REPR_3 | REPR_4 |
|---|---|---|---|---|
| CONCEPT...CUCUMBER | 0.64 (0.35) | 0.19 (0.1) | 0.41 (0.2) | 0.46 (0.25) |
| CONCEPT...TEA_CUP | 0.69 (0.4) | 0.44 (0.25) | 0.2 (0.08) | 0.32 (0.19) |
| CONCEPT...KETTLE | 0.82 (0.49) | 0.54 (0.3) | 0.39 (0.24) | 0.17 (0.08) |

**Table 4.14:** Distances between representatives and concepts: Tanimoto-KL (numbers) and Sørenson-$L^2$ (numbers in braces).

### 4.4.6 Fulfillment of World Modeling Requirements

In Sec. 2.2.2 we have defined formal requirements, which have to be fulfilled in order to obtain practically useful world modeling. Hereby, we will discuss compliance of $PM^2$ to these requirements.

- *correctness* – the contents of the dynamic world model is matching the real scene with an error of sensors uncertainty (e.g. about 0.1 m spatial uncertainty for entities at 1.5 m distance for ARMAR-III). The prior knowledge is pre-defined in order to reflect the environment sufficiently for all planned tasks. The functionality of the system is found to be sufficient for the ARMAR robot and the NEST system under test conditions;

- *minimality* – an information reduction is performed by separating dynamic information and prior knowledge and by aging mechanisms.
- *universality* – the $PM^2$ platform is able to model probability distributions in form of Gaussian mixtures or particles, covering all required information types. Distributions of all types are processed in a unified fashion within one general information management framework. Any relations between representatives are modeled as multiple semantic networks. Due to the abstract description nature, the Progressive Mapping allows handling of unknown entities;
- *semanticity* – the semantic modeling consists of two important parts: the attributes global scheme and the concepts ontology. The attributes scheme defines which attributes, units and coordinates are standard to the modeling system. The concepts ontology allows for meaningful entities handling and semantic analysis of the scene at hand;
- *robustness and efficiency* – the fulfillment of correctness and minimality requirements combined with multi-threading processing allows for real-time functioning of the NEST system on average computer systems and on the ARMAR-III robot;
- *dispatch* – a central disposition of the world modeling subsystem within NEST and ARMAR guarantees free model-complied information exchange between sub-modules;
- *clarity* – the data interchange between sub-modules is performed in XML format understandable both for humans and machines. Moreover, these data streams can be at any time visualized in 3D on dedicated monitor stations.

# 5

# Conclusion

## 5.1 Summary

Within the scope of our work, we have analyzed existing and proposed new theoretical approaches for world modeling, as well as implemented and tested them in practice. This includes areas of dynamic and prior knowledge modeling, information association, fusion and management, qualitative and quantitative information analysis. Namely, we have proposed a dynamic modeling of arbitrary entities (even unknown and not deterministically classifiable). In this environment modeling, information is represented in form of marginal and joint degree-of-belief (DoB) distributions (including mixed joints of discrete and continuous distributions) assigned to progressive containers created within the object-oriented paradigm. The DoB representation allows for uncertainty incorporation and powerful Bayesian fusion mechanisms. We have discussed possibilities for qualitative scene analysis (e.g. temporal situation assessment over semantic networks). On the other hand, the proposed information entropy framework has allowed for quantitative information assessment (e.g. information sufficiency relative to given tasks). Moreover, we have researched possibilities of prior knowledge modeling and employment of persistent information in dynamic models. This allows for probabilistic classification with subsequent weighted update with prior knowledge. Finally, we have systematically analyzed possibilities of physical parameters employment, e.g. sensor acceptance and task-required precision. To this end, we have introduced the least discernible quantum for entropy calculus unification as well as for particle filter resampling. We have employed sensor- and task-dependent probe functions for generalized functions calculus applied to distributions difference metric problem and Gaussian mixture reduction. In addition to theoretical analysis, we have implemented and experimentally evaluated the proposed methods and approaches on two hardware set-ups: ARMAR-III and NEST. A complete list of contributions is presented in Sec. 1.4.2.

## 5.2 Outlook

Due to complexity of the topic under consideration, there is still much work until intelligent autonomous robots can approach the desired level of cognition and situation awareness. Therefore, in the following section we will mention yet unsolved issues related to the current work.

### 5.2.1 Open Issues

*Bestimmtheit Assessment* The Bestimmtheit calculation can be extended for taking into account relations between representatives and among attributes. For example, Bestimmtheit assessment for joint DoB distributions requires a generalization of our entropy framework.

*Weak Distance* There are several open questions in weak distance employment for GM reduction and tracking. First, it is still not clear why the WC algorithm leads to relatively large WSD losses. Second, there is no estimation on how discreteness of the information representation limits the WD algorithm due to arising systematic effects. Third, it is necessary to find conditions and applications that would benefit from the proposed algorithms the most and what optimal probe function sets correspond the given applications.

# Bibliography


[Afo08] AFONSO, Manya: Particle Filter and Extended Kalman Filter for Nonlinear Estimation: A Comparative Study, Tech. Rep., Instituto de Telecomunicações, Instituto Superior Técnico (2008)

[Ahl02] AHLRICHS, Ulrike: *Wissensbasierte Szenenexploration auf der Basis erlernter Analysestrategien,* Ph.D. thesis, University Erlangen-Nürnberg, Logos Verlag Berlin (2002)

[Ahn11] AHN, Ho Seok; LEE, Dong-Wook; CHOI, Dongwoon; LEE, Duk Yeon; HUR, Man Hong; LEE, Hogil and SHON, Woong Hee: Development of an Android for Singing with Facial Expression, in: *IECON 2011 - 37th Annual Conference on IEEE Industrial Electronics Society,* IEEE (2011), pp. 104–109

[Ahr05] AHRENDT, P.: The Multivariate Gaussian Probability Distribution, Tech. Rep., IMM, Technical University of Denmark (2005)

[Amb00] AMBROSE, Robert; ALDRIDGE, H.; ASKEW, R.S.; BURRIDGE, R.; BLUETHMAN, W.; DIFTLER, M.A.; LOVCHIK, C.; MAGRUDER, D. and REHNMARK, F.: ROBONAUT: NASA's Space Humanoid. *IEEE Intelligent Systems Journal* (2000), vol. 15(4):pp. 57–63

[And95] ANDERSON, John Eric: *Constraint-Directed Improvisation For Everyday Activities,* Ph.D. thesis, University of Manitoba (1995)

[Ang92] ANGELOPOULOU, Elli; TSAI-HONG, Hong and WU, Angela: World Model Representations for Mobile Robots, in: *Proceedings of the IEEE Intelligent Vehicles Symposium*, IEEE (1992), pp. 293–297

[Ark98] ARKIN, Ronald C.: *Behavior-Based Robotics*, The MIT Press (1998)

[Arr98] ARRAS, K.O. and VESTLI, S.J.: Hybrid, high-precision localization for the mail distributing mobile robot system, in: *Proceedings of the IEEE International Conference on Robotics and Automation*, vol. 4, IEEE (1998), pp. 3129–3134

[Asf06] ASFOUR, T.; REGENSTEIN, K.; AZAD, P.; SCHRÖDER, J. and DILLMANN, R.: ARMAR-III: A Humanoid Platform for Perception-Action Integration, in: *2nd International Workshop on Human-Centered Robotic Systems (HCRS)* (2006)

[Asf08] ASFOUR, T.; WELKE, K.; AZAD, P.; UDE, A. and DILLMANN, R.: The Karlsruhe Humanoid Head, in: *Proceedings of the 8th IEEE-RAS International Conference on Humanoid Robots*, IEEE (2008), pp. 447–453

[Asi48] ASIMOV, Isaac: Runaround, in: *Astounding Science Fiction*, Street & Smith (1948)

[Aus00] AUSTIN, D.J. and JENSFELT, P.: Using multiple Gaussian hypotheses to represent probability distributions for mobile robot localization, in: *Proceedings of the IEEE International Conference on Robotics and Automation*, vol. 2, IEEE (2000), pp. 1036–1041

[Bal98] BALCH, T. and ARKIN, R.C: Behavior-based formation control for multirobot teams. *IEEE Transactions on Robotics and Automation* (1998), vol. 14:pp. 926–939

[Ban08] BANKO, Michele and ETZIONI, Oren: The Tradeoffs Between Open and Traditional Relation Extraction, in: *Proceedings of ACL-08: HLT*, Association for Computational Linguistics, Columbus, Ohio (2008), pp. 28–36, URL `http://www.aclweb.org/anthology/P/P08/P08-1004`

[Bat04] BATEMAN, John and FARRAR, Scott: Modelling models of robot navigation using formal spatial ontology, in: *Proceedings of Spatial Cognition*, Springer (2004), pp. 366–389

[Bau09a] BAUER, A.: Probabilistic reasoning on object occurrence in complex scenes, in: *Image and Signal Processing for Remote Sensing XV, Proceedings of SPIE*, vol. 74770A (2009)

[Bau09b] BAUER, A.; EMTER, T.; VAGTS, H. and BEYERER, J.: Object-oriented world model for surveillance systems, in: P. Elsner (Editor) *Future Security. 4th Security Research Conference Karlsruhe*, Fraunhofer-Gesellschaft, Fraunhofer Verlag (2009), pp. 339–345

[Bau10a] BAUM, Marcus; GHEŢA, Ioana; BELKIN, Andrey; BEYERER, Jürgen and HANEBECK, Uwe D.: Data Association in a World Model for Autonomous Systems, in: *Proceedings of the 2010 IEEE International Conference on Multisensor Fusion and Integration for Intelligent Systems*, IEEE, Omnipress, Fort Douglas, University of Utah, Salt Lake City, USA (2010), pp. 187–192, URL `http://isas.uka.de/Publikationen/MFI10_BaumGheta.pdf`

[Bau10b] BAUM, Marcus; GHEŢA, Ioana; BELKIN, Andrey; BEYERER, Jürgen and HANEBECK, Uwe D.: Three Pillar Information Management System for Modeling the Environment of Autonomous Systems, in: *Proceedings of the 2010 IEEE International Conference on Virtual Environments, Human-Computer Interfaces and Measurement Systems (VECIMS)*, IEEE, Taranto, Italy (2010), pp. 12–17, URL `http://isas.uka.de/Publikationen/VECIMS10_GhetaBaum.pdf`

[Bau13] BAUER, Alexander: *Probabilistische Szenenmodelle für die Luftbildauswertung*, Ph.D. thesis, Karlsruher Institut für Technologie (KIT), KIT Scientific Publishing (2013)

[Bay08] BAY, Herbert; ESS, Andreas; TUYTELAARS, Tinne and VAN GOOL, Luc: Speeded-Up Robust Features (SURF). *Computer Vision and Image Understanding* (2008), vol. 110(3):pp. 346–359, URL `http://dx.doi.org/10.1016/j.cviu.2007.09.014`

[Bee10] BEETZ, Michael; MÖSENLECHNER, Lorenz and TENORTH, Moritz: CRAM – A Cognitive Robot Abstract Machine for Everyday Manipulation in Human Environments, in: *Proceedings of the IEEE/RSJ International Conference on Intelligent Robots and Systems*, IEEE, Taipei, Taiwan (2010), pp. 1012–1017

[Bel10] BELKIN, Andrey: Object-Oriented World Modelling for Autonomous Systems, in: Jürgen Beyerer and Marco Huber (Editors) *Proceedings of the 2009 Joint Workshop of Fraunhofer IOSB and Institute for Anthropomatics, Vision and Fusion Laboratory*, vol. 4 of *Karlsruher Schriften zur Anthropomatik*, Vision and Fusion Laboratory (IES), Institute for Anthropomatics (IFA), Karlsruhe Institute for Technology (KIT), Fraunhofer Institute of Optronics, System Technologies and Image Exploitation (IOSB), KIT Scientific Publishing, La Bresse, France (2010), pp. 231–246, URL http://digbib.ubka.uni-karlsruhe.de/volltexte/documents/1252397

[Bel11] BELKIN, Andrey: Information Management in World Modeling, in: Jürgen Beyerer and Marco Huber (Editors) *Proceedings of the 2010 Joint Workshop of Fraunhofer IOSB and Institute for Anthropomatics, Vision and Fusion Laboratory*, vol. 7 of *Karlsruher Schriften zur Anthropomatik*, Vision and Fusion Laboratory (IES), Institute for Anthropomatics (IFA), Karlsruhe Institute for Technology (KIT), Fraunhofer Institute of Optronics, System Technologies and Image Exploitation (IOSB), KIT Scientific Publishing, La Bresse, France (2011), pp. 187–199, URL http://digbib.ubka.uni-karlsruhe.de/volltexte/documents/1625195

[Bel12a] BELKIN, Andrey: Dynamic World Modeling with Prior Knowledge Matching for Autonomous Systems, in: Alexey Pak Jürgen Beyerer (Editor) *Proceedings of the 2011 Joint Workshop of Fraunhofer IOSB and Institute for Anthropomatics, Vision and Fusion Laboratory*, vol. 11 of *Karlsruher Schriften zur Anthropomatik*, Vision and Fusion Laboratory (IES), Institute for Anthropomatics (IFA), Karlsruhe Institute for Technology (KIT), Fraunhofer Institute of Optronics, System Technologies and Image Exploitation (IOSB), KIT Scientific Publishing, Triberg, Germany (2012), pp. 175–184, URL http://digbib.ubka.uni-karlsruhe.de/volltexte/documents/2197120

[Bel12b] Belkin, Andrey and Beyerer, Jürgen: Information Entropy and Structural Metrics Based Estimation of Situations as a Basis for Situation Awareness and Decision Support, in: *2012 IEEE International Multi-Disciplinary Conference on Cognitive Methods in Situation Awareness and Decision Support (CogSIMA 2012)*, CFP12COH-CDR, IEEE, CUBRC, New Orleans, LA, USA (2012), pp. 111–116, URL http://cogsima2012.org

[Bel12c] Belkin, Andrey and Beyerer, Jürgen: *Intelligent Robotics and Applications*, vol. 7508 of *Lecture Notes in Artificial Intelligence*, chap. Prior Knowledge Employment Based on the K-L and Tanimoto Distances Matching for Intelligent Autonomous Robots, Springer, Concordia University, Montreal, Quebec, Canada (2012), pp. 171–180, the 5th International Conference on Intelligent Robotics and Applications

[Bel12d] Belkin, Andrey; Kuwertz, Achim; Fischer, Yvonne and Beyerer, Jürgen: *Innovative Information Systems Modelling Techniques*, vol. 1, chap. World Modeling for Autonomous Systems, InTech – Open Access Publisher (2012), pp. 135–156, URL http://www.intechopen.com/books/innovative-information-systems-modelling-techniques

[Ben06] Benenson, R.; Petti, S.; Fraichard, T. and Parent, M.: Integrating Perception and Planning for Autonomous Navigation of Urban Vehicles, in: *IEEE/RSJ International Conference on Intelligent Robots and Systems*, IEEE (2006), pp. 98–104

[Biz09] Bizer, Chris; Auer, Sören; Kobilarov, Georgi; Lehmann, Jens; Becker, Christian and Hellmann, Sebastian: DBpedia – Querying Wikipedia like a Database and An Interlinking-Hub in the Web of Data (2009), URL http://www4.wiwiss.fu-berlin.de/bizer/pub/WikiMediaDevMeeting-DBpedia-Talk.pdf, querying Wikipedia Like a Database (4/4/2009) FU Berlin, Universität Leipzig

[Boo93] BOOCH, Grady: *Object-Oriented Analysis and Design with Applications*, Addison-Wesley Professional, 2nd edn. (1993)

[Bou96] BOUTILIER, Craig: Planning, learning and coordination in multiagent decision processes, in: *Proceedings of the Conference on Theoretical Aspects of Rationality and Knowledge*, Morgan Kaufmann Publishers Inc. (1996), pp. 195–210

[BS88] BAR-SHALOM, Yaakov and FORTMANN, Thomas E.: *Tracking and data association*, Academic Press (1988)

[BS01] BAR-SHALOM, Yaakov; LI, X. Rong and KIRUBARAJAN, Thiagalingam: *Estimation with Applications to Tracking and Navigation*, Wiley-Interscience, 1st edn. (2001), URL `http://www.worldcat.org/isbn/047141655X`

[BS09] BAR-SHALOM, F., Y.AND Daum and HUANG, J.: The probabilistic data association filter. *IEEE Control Systems* (2009), vol. 29(6):pp. 82–100

[Bue07] BUEHLER, Martin; IAGNEMMA, Karl and SINGH, Sanjiv (Editors): *The 2005 DARPA Grand Challenge*, vol. 36 of *Springer Tracts in Advanced Robotics*, Springer (2007)

[Bue10] BUEHLER, Martin; IAGNEMMA, Karl and SINGH, Sanjiv (Editors): *The DARPA Urban Challenge*, vol. 56 of *Springer Tracts in Advanced Robotics*, Springer (2010)

[Bur02] BURNHAM, Kenneth P. and ANDERSON, David R.: *Model Selection and Multimodel Inference: A Practical Information-Theoretic Approach*, Springer (2002)

[Cau00] CAULFIELD, H. John and JOHNSON, John L.: Artificial perception and consciousness, in: *Proceedings of the Sixth International Conference on Education and Training in Optics and Photonics*, vol. 3831 (2000)

[Cha07] CHA, Sung-Hyuk: Comprehensive Survey on Distance/Similarity Measures between Probability Density Functions. *International Journal of Mathematical Models and Methods in Applied Sciences* (2007), vol. 1(4):pp. 300–307, URL http://www.gly.fsu.edu/~parker/geostats/Cha.pdf

[Cha11] CHALLA, Subhash; MORELANDE, Mark R.; MUŠICKI, Darko and EVANS, Robin J.: *Fundamentals of OBJECT TRACKING*, Cambridge University Press (2011)

[Che03] CHEN, Zhe: Bayesian Filtering: From Kalman Filters to Particle Filters, and Beyond, Tech. Rep., Communications Research Laboratory, McMaster University (2003)

[Chk05] CHKLOVSKI, Timothy and GIL, Yolanda: Improving the design of intelligent acquisition interfaces for collecting world knowledge from web contributors, in: *Proceedings of the 3rd international conference on Knowledge capture*, K-CAP '05, ACM, New York, NY, USA (2005), pp. 35–42, URL http://doi.acm.org/10.1145/1088622.1088630

[Chu08] CHUNG, Yong-Joo: *Intelligent Data Engineering and Automated Learning*, vol. 5326/2008 of *LNCS*, chap. Using Kullback-Leibler Distance in Determining the Classes for the Heart Sound Signal Classification, Springer (2008), pp. 49–56

[Cos08] COSTA, Paulo Cesar; D'AMATO, Claudia; FANIZZI, Nicola; LASKEY, Kathryn B.; LASKEY, Kenneth J.; LUKASIEWICZ, Thomas; NICKLES, Matthias and POOL, Michael (Editors): *Uncertainty Reasoning for the Semantic Web I*, vol. 5327 of *Lecture Notes in Artificial Intelligence*, Springer-Verlag (2008), URL http://dx.doi.org/10.1007/978-3-540-89765-1_6

[Cov91] COVER, Thomas M. and THOMAS, Joy A.: *Elements of Information Theory*, Wiley-Interscience (1991)

[Cro11] CROUSE, David F.; WILLETT, Peter; PATTIPATI, Krishna and SVENSSON, Lennart: A Look At Gaussian Mixture Reduction Algorithms, in: *Proceedings of the 14th International Conference on Information Fusion*, Chicago, IL (2011)

[DAR08] DARPA: Legged Squad Support System (LS3), Broad Agency Announcement (BAA), Defense Advanced Research Projects Agency DARPA/TACTICAL TECHNOLOGY OFFICE (TTO) 3701 N. Fairfax Drive Arlington, VA 22203-1714 (2008)

[DAR10] DARPA: Autonomous Robot Manipulation (ARM) (2010), URL `http://thearmrobot.com/`

[Das08] DAS, Subrata: *High-Level Data Fusion*, Art (2008)

[Dav04] DAVISON, A.J.; CID, Y. González and KITA, N.: Real-Time 3D SLAM with Wide-Angle Vision, in: *Proceedings of the IFAC Symposium on Intelligent Autonomous Vehicles* (2004)

[Dem95] DEMENTHON, Daniel F. and DAVIS, Larry S.: Model-based Object Pose in 25 Lines of Code. *International Journal of Computer Vision* (1995), vol. 15(1-2):pp. 123–141, URL `http://dx.doi.org/10.1007/BF01450852`

[Deu08] DEUTSCHE FORSCHUNGSGEMEINSCHAFT (DFG): Collaborative Research Center (SFB) 588 (2008), URL `http://www.sfb588.uni-karlsruhe.de`

[Dhi02] DHILLON, Inderjit S.; MALLELA, Subramanyam and KUMAR, Rahul: Enhanced word clustering for hierarchical text classification, in: *Proceedings of the eighth ACM SIGKDD international conference on Knowledge discovery and data mining*, ACM (2002), pp. 191–200

[Dic94] DICKMANNS, E.D.; BEHRINGER, R.; DICKMANNS, D.; HILDEBRANDT, T.; MAURER, M.; THOMANEK, F. and SCHIEHLEN, J.: The Seeing Passenger Car 'Va-MoRs-P', in: *Proceedings of the Intelligent Vehicles '94 Symposium* (1994), pp. 68–73

[Die12] DIETRICH, André; ZUG, Sebastian and KAISER, Jörg: *Computer Safety, Reliability, and Security*, vol. 7613 of *Lecture Notes in Computer Science*, chap. Towards Artificial Perception, Springer (2012), pp. 466–476

[Dis01] DISSANAYAKE, M.W.M.G.; NEWMAN, P.; CLARK, S.; DURRANT-WHYTE, H.F. and CSORBA, M.: A Solution to the Simultaneous Localisation and Map Building (SLAM) Problem. *IEEE Transactions on Robotics and Automation* (2001), vol. 17:pp. 229–241

[Dob14] DOBRUSHKIN, Vladimir A.: *Applied Differential Equations: An Introduction*, Chapman and Hall/CRC (2014)

[Dor11] DORIGO, M.; FLOREANO, D.; GAMBARDELLA, L. M.; MONDADA, F.; S. NOLFI, T. Baaboura; BIRATTARI, M.; BONANI, M.; BRAMBILLA, M.; BRUTSCHY, A.; BURNIER, D.; A. CAMPO, A. L. Christensen; DECUGNIÈRE, A.; CARO, G. Di; DUCATELLE, F.; FERRANTE, E.; A. FÖRSTER, J. Martinez Gonzales; GUZZI, J.; LONGCHAMP, V.; MAGNENAT, S.; N. MATHEWS, M. Montes de Oca; O'GRADY, R.; PINCIROLI, C.; PINI, G.; RÉTORNAZ, P.; ROBERTS, J.; V. SPERATI, T. Stirling; STRANIERI, A.; STÜTZLE, T.; TRIANNI, V.; TUCI, E.; TURGUT, A. E. and VAUSSARD, F.: Swarmanoid: a novel concept for the study of heterogeneous robotic swarms, Tech. Rep., Institut de Recherches Interdisciplinaires et de Développements en Intelligence Artificielle, Université Libre de Bruxelles (2011)

[Emt07] EMTER, Thomas; MONARI, Eduardo; FREY, Christian; MÜLLER, Thomas; KUNTZE, Helge-Björn; LAUBENHEIMER, Astrid and MÜLLER, Markus: AMROS – an Autonomous Mobile Robotic System for Multisensor Surveillance of Real Estates, in: *Proceedings of the Future Security, 2nd Security Conference*, Universitätsverlag Karlsruhe (2007), pp. 151–154

[Emt10] EMTER, T.: Probabilistic localization and mapping for mobile robots, in: J. Beyerer and M. Huber (Editors) *Proceedings of the Joint Workshop of Fraunhofer IOSB and Institute for Anthropomatics, Vision and Fusion Laboratory 2010*, vol. 7 of *Karlsruher Schriften zur Anthropomatik*, KIT Scientific Publishing (2010), pp. 91–106

[End95] ENDSLEY, Mica: Toward a theory of situation awareness in dynamic systems. *Human Factors* (1995), vol. 37(1):pp. 32–64

[Fel71] FELLER, William: *An Introduction to Probability and Its Applications,* vol. 2, Wiley (1971)

[Fis11] FISCHER, Y.; BAUER, A. and BEYERER, J.: A conceptual framework for automatic situation assessment, in: *2011 IEEE First International Multi-Disciplinary Conference on Cognitive Methods in Situation Awareness and Decision Support (CogSIMA),* IEEE (2011), pp. 234–239

[Fis15] FISCHER, Yvonne: *Wissensbasierte probabilistische Modellierung für die Situationsanalyse am Beispiel der maritimen Überwachung,* Ph.D. thesis, Karlsruher Institut für Technologie (KIT), KIT Scientific Publishing (2015)

[Fit08] FITZPATRICK, P.; METTA, G. and NATALE, L.: Towards long-lived robot genes. *Robotics and Autonomous Systems* (2008), vol. 56:pp. 29–45

[Fra08] FRAUNHOFER INSTITUTE OF OPTRONICS, SYSTEM TECHNOLOGIES AND IMAGE EXPLOITATION (IOSB): (2008), URL `http://www.iosb.fraunhofer.de`

[Fri98] FRIEDRICH, Holger: *Interaktive Programmierung von Manipulationssequenzen,* Ph.D. thesis, Karlsruhe University (TH), GCA-Verlag (1998)

[Fur10] FURDA, Andrei and VLACIC, Ljubo: An Object-Oriented Design of a World Model for Autonomous City Vehicles, in: *IEEE Intelligent Vehicles Symposium (IV),* IEEE (2010), pp. 1054–1059

[Gan02] GANDON, Fabien L.: Ontology Engineering: a Survey and a Return on Experience, Tech. Rep. 4396, INSTITUT DE RECHERCHE EN INFORMATIQUE ET AUTOMATIQUE (2002)

[Gel64] GELFAND, I.M. and SHILOV, G.E.: *Generalized Functions. Vol I: Properties and operations,* Academic Press, Boston, USA (1964), translated from Russian

[Gel74] GELB, Arthur: *Applied Optimal Estimation*, The MIT Press (1974), URL http://www.worldcat.org/isbn/0262570483

[Ger03] GERKEY, B.; VAUGHAN, R. T. and HOWARD, A.: The Player/Stage Project: Tools for multi-robot and distributed sensor systems, in: *Proceedings of the 11th International Conference on Advanced Robotics* (2003), pp. 317–323

[Ghe08] GHEŢA, Ioana; HEIZMANN, Michael and BEYERER, Jürgen: Object oriented environment model for autonomous systems, in: *SWIFT 2008 – SkövdeWorkshop on Information Fusion* (2008), pp. 9–12

[Ghe10] GHEŢA, Ioana; HEIZMANN, Michael; BELKIN, Andrey and BEYERER, Jürgen: *KI 2010*, vol. 6359 of *Lecture Notes in Artificial Intelligence*, chap. World Modeling for Autonomous Systems, Springer, Karlsruhe, Germany (2010), pp. 176–183, URL http://www.springerlink.com/content/322qw21017t56x31/fulltext.pdf, 33rd Annual German Conference on Artificial Intelligence (KI 2010)

[Gib02] GIBBS, Alison L. and SU, Francis Edward: On Choosing and Bounding Probability Metrics. *International Statistical Review* (2002), vol. 70(3):pp. 419–435

[Gli06] GLINTON, R.; GIAMPAPA, J. and SYCARA, K.: A Markov Random Field Model of Context for High-Level Information Fusion, in: *Proceedings of the 9th International Conference on Information Fusion*, IEEE (2006), pp. 1–8

[Go05] GO, Young Cheol and SOHN, Joo-Chan: Context modeling for intelligent robot services using rule and ontology, in: *The 7th International Conference on Advanced Communication Technology*, vol. 2, Springer (2005), pp. 813–816

[Göh07] GÖHRING, Daniel and BURKHARD, Hans-Dieter: Cooperative World Modeling in Dynamic Multi-Robot Environments. *Fundamenta Informaticae* (2007), vol. 75(1-4):pp. 281–294

[Göh08] GÖHRING, Daniel; MELLMANN, Heinrich; GERASYMOVA, Kataryna and BURKHARD, Hans-Dieter: Constraint Based World Modeling. *Fundamenta Informaticae* (2008), vol. 85(1-4):pp. 123–137

[Göh09] GÖHRING, Daniel: *Constraint Based World Modeling for Multi Agent Systems in Dynamic Environments*, Ph.D. thesis, Mathematisch-Naturwissenschaftlichen Fakultät II Humboldt-Universität zu Berlin, Springer-Verlag (2009)

[Gri02] GRIFFITHS, Gwyn and GRIFFITHS, Griffiths: *Technology and Applications of Autonomous Underwater Vehicles*, Routledge Chapman & Hall (2002)

[Gro01] GROEN, F.C.A.; ROODHART, J.; SPAAN, M.; DONKERVOORT, R. and VLASSIS, N.: A distributed world model for robot soccer that supports the development of team skills, in: *Proceedings of the 13th Belgian-Dutch Conference on Artificial Intelligence* (2001), pp. 389–396

[Grü07] GRÜNWALD, Peter D.: *The Minimum Description Length Principle*, MIT Press (2007)

[Haa04] HAASCH, A.; HOHENNER, S.; HÜUWEL, S.; KLEINEHAGENBROCK, M.; LANG, S.; TOPTSIS, I.; FINK, G. A.; FRITSCH, J.; WREDE, B. and SAGERER, G.: BIRON – The Bielefeld Robot Companion, in: *International Workshop on Advances in Service Robotics*, Fraunhofer IRB Verlag (2004), pp. 27–32

[Hal04] HALL, David L. and MCMULLEN, Sonya A.H. (Editors): *Mathematical Techniques in Multisensor Data Fusion*, Artech House, Inc, 2nd edn. (2004)

[Han08] HANEBECK, U.D. and KLUMPP, V.: Localized Cumulative Distributions and a multivariate generalization of the Cramér-von Mises distance, in: *International Conference on Multisensor Fusion and Integration for Intelligent Systems*, IEEE (2008), pp. 33–39

[Har88] HARRIS, Chris and STEPHENS, Mike: A combined corner and edge detector, in: *Proceedings of the Fourth Alvey Vision Conference* (1988), pp. 147–151

[Har04] HARLE, R. K. and HOPPER, A.: Dynamic World Models from Raytracing, in: *Proceedings of the Second IEEE International Conference on Pervasive Computing and Communications (PerCom'04)*, IEEE (2004), pp. 55–64

[Haz01] HAZEWINKEL, Michiel (Editor): *Encyclopedia of Mathematics*, chap. Wasserstein metric, Springer (2001), URL `http://www.encyclopediaofmath.org/index.php?title=Wasserstein_metric&oldid=32292`

[Her07] HERSHEY, John R. and OLSEN, Peder A.: Approximating the Kullback Leibler Divergence Between Gaussian Mixture Models, in: *International Conference on Acoustics, Speech and Signal Processing (ICASSP)*, vol. 4, IEEE (2007), pp. IV–317 – IV–320

[Hig97] HIGUCHI, Tomoyuki: Monte Carlo filter using the genetic algorithm operators. *Journal of Statistical Computation and Simulation* (1997), vol. 59(1):pp. 1–23

[Hon07] HONDA MOTOR CO., LTD.: ASIMO, The Honda HUMANOID ROBOT, Technical Information (2007)

[Hor09] HORRIDGE, P. and MASKELL, S.: Searching for, initiating and tracking multiple targets using existence probabilities, in: *FUSION '09. 12th International Conference on Information Fusion*, IEEE (2009), pp. 611–617

[Hsi03] HSIAO, Kai-Yuh; MAVRIDIS, Nikolaos and ROY, Deb: Coupling perception and simulation: steps towards conversational robotics, in: *Proceedings of the IEEE/RSJ International Conference on Intelligent Robots and Systems*, vol. 1, IEEE (2003), pp. 928–933

[Hub08] HUBER, Marco F. and HANEBECK, Uwe D.: Progressive Gaussian Mixture Reduction, in: *Proceedings of the 11th International Conference on Information Fusion (Fusion)*, Cologne, Germany (2008)

[Hub09] HUBER, Marco: *Probabilistic Framework for Sensor Management*, Ph.D. thesis, Universität Karlsruhe (TH), Universitätsverlag Karlsruhe (2009)

[IBM01] IBM: Autonomic Computing: IBM's Perspective on the State of Information Technology, Manifesto (2001), URL http://www.research.ibm.com/autonomic/manifesto/

[Iso01] ISODA, S.: Object-Oriented World-Modeling Revisited. *Journal of Systems and Software* (2001), vol. 59(2):pp. 153–162

[Jac89] JACKSON, Donald A.; SOMERS, Keith M. and HARVEY, Harold H.: Similarity Coefficients: Measures of Co-occurrence and Assosication or Simply Measures of Occurrence? *The American Naturalist* (1989), vol. 133:pp. 436–453

[Jai12] JAIN, Dominik: *Probabilistic Cognition for Technical Systems: Statistical Relational Models for High-Level Knowledge Representation, Learning and Reasoning*, Ph.D. thesis, Technische Universität München (2012)

[Jäk11] JÄKEL, R.; SCHMIDT-ROHR, S.R.; RÜHL, S.W.; KASPER, A.; XUE, Z. and DILLMANN, R.: Learning of Planning Models for Dexterous Manipulation Based on Human Demonstrations. *International Journal of Social Robotics* (2011), vol. 4:pp. 437–448

[Jam01] JAMIL, N.; LQBAL, S. and IQBAL, N.: Face recognition using neural networks, in: *Proceedings of the IEEE International Multi Topic Conference. INMIC 2001. Technology for the 21st Century*, IEEE (2001), pp. 277–281

[Jen07] JENSEN, Jesper Højvang; ELLIS, Daniel P.W.; CHRISTENSEN, Mads G. and JENSEN, Søren Holdt: Evaluation Distance Measures Between Gaussian Mixture Models of MFCCs, in: *ISMIR 2007: Proceedings of the 8th International Conference on Music Information Retrieval*, vol. 2, Austrian Computer Society (2007), pp. 107–108

[Kai11] KAISER, Peter; BERENSON, Dmitry; VAHRENKAMP, Nikolaus; ASFOUR, Tamim; DILLMANN, Rüdiger; SRINIVASA, Siddhartha and PLANNING, Abstract: Constellation - An Algorithm for Finding Robot Configurations that Satisfy Multiple Constraints, in: *IEEE/RSJ International Conference on Intelligent Robots and Systems (IROS)*, IEEE (2011), pp. 436–443

[Kal60] Kalman, R. E.: A New Approach to Linear Filtering and Prediction Problems. *Transactions of the ASME – Journal of Basic Engineering* (1960), vol. 82(1):pp. 35–45

[Kan08] Kaneko, Kenji; Harada, Kensuke; Kanehiro, Fumio; Miyamori, Go and Akachi, Kazuhiko: Humanoid Robot HRP-3, in: *IEEE/RSJ International Conference on Intelligent Robots and Systems*, IEEE (2008)

[Kar06] Karakoç, Emre; Cherkasov, Artem and Sahinalp, Süleyman Cenk: Distance based algorithms for small biomolecule classification and structural similarity search, in: *ISMB (Supplement of Bioinformatics)'06*, Oxford University Press (2006), pp. 243–251

[Kas13] Kasper, Alexander: *Szenen- und Objektmodellierung für Servicerroboter*, Ph.D. thesis, Karlsruhe Institute of Technology (2013)

[KB05] Krieg-Bruckner, B.; Frese, U.; Luttich, K.; Mandel, C.; Mossakowski, T. and Ross, R. J.: *Spatial Cognition IV*, chap. Specification of an ontology for route graphs, Lecture Notes in Computer Science, Springer-Verlag (2005), pp. 390–412

[Kep03] Kephart, Jeffrey O. and Chess, David M.: The Vision of Autonomic Computing. *Computer* (2003), vol. 36(1):pp. 41–50

[Kit96] Kitagawa, Genshiro: Monte Carlo Filter and Smoother for Non-Gaussian Nonlinear State Space Models. *Journal of Computational and Graphical Statistics* (1996), vol. 5(1):pp. 1–25

[Kok02] Kok, J.R. and Vlassis, N.: Mutual modeling of teammate behavior, Tech. Rep., Computer Science Institute, University of Amsterdam, Amsterdam, Netherlands (2002)

[Kot03] Kotecha, Jayesh. H. and Djurić, Petar M.: Gaussian Sum Particle Filtering. *IEEE Transactions on Signal Processing* (2003), vol. 51(10):pp. 2602–2612

[Kri13] Kriegel, Simon; Brucker, Manuel; Marton, Zoltan-Csaba; Bodenmüller, Tim and Suppa, Michael: Combining object modeling and recognition for active scene exploration, in: *IEEE/RSJ*

*International Conference on Intelligent Robots and Systems*, IEEE (2013), pp. 2384–2391

[Küh10] KÜHN, Benjamin; BELKIN, Andrey; SWERDLOW, Alexej; MACHMER, Timo; BEYERER, Jürgen Beyerer and KROSCHEL, Kristian: Knowledge-Driven Opto-Acoustic Scene Analysis based on an Object-Oriented World Modeling approach for Humanoid Robots, in: *Proceedings of the 41st International Symposium on Robotics and the 6th German Conference on Robotics (ISR/ROBOTIK 2010)*, ITG (VDE) Information Technology Society of VDE, VDMA Robotik + Automation, IFR International Federation of Robotics, DGR German Association on Robotics, VDE Verlag GmbH, Munich, Germany (2010), URL `http://opasca.org/publications/2010_06_isr-robotik2010.pdf`

[Küh15] KÜHN, Benjamin: *Interessengetriebene audiovisuelle Szenenexploration*, Ph.D. thesis, Karlsruher Institut für Technologie (KIT), KIT Scientific Publishing (2015)

[KUK08] KUKA ROBOTER GMBH: KUKA robot hand (2008), URL `http://www.kuka-robotics.com`

[Kul51] KULLBACK, S. and LEIBLER, R. A.: On information and sufficiency. *Annals of Mathematical Statistics* (1951), vol. 22:pp. 49–86

[Kus08] KUSUDA, Yoshihiro: Toyota's violin-playing robot. *Industrial Robot: An International Journal* (2008), vol. 35(6):pp. 504–506

[Kuw13] KUWERTZ, Achim: Extending Object-Oriented World Modeling for Adaptive Open-World Modeling, in: Jürgen Beyerer and Alexey Pak (Editors) *Proceedings of the 2012 Joint Workshop of Fraunhofer IOSB and Institute for Anthropomatics, Vision and Fusion Laboratory*, vol. 13 of *Karlsruher Schriften zur Anthropomatik*, Vision and Fusion Laboratory (IES), Institute for Anthropomatics (IFA), Karlsruhe Institute for Technology (KIT), Fraunhofer Institute of Optronics, System Technologies and Image Exploitation (IOSB), KIT Scientific Publishing (2013)

[Lah08] LAHR, D.F. and HONG, D.W.: The Development of CHARLI: A Linear Actuated Powered Full Size Humanoid Robot, in: *Proceedings URAI 2008*, Seoul, Korea (2008)

[Leh11] LEHRSTUHL FÜR INTERAKTIVE ECHTZEITSYSTEME (IES): (2011), URL `http://ies.anthropomatik.kit.edu/`

[Len90] LENAT, Douglas B. and GUHA, R. V.: *Building Large Knowledge-Based Systems: Representation and Inference in the Cyc Project*, Addison-Wesley Pub (1990)

[Leo02] LEONARD, J. J. and DURRANT-WHYTE, H. F.: Mobile robot localization by tracking geometric beacons. *Robotics and Automation, IEEE Transactions on* (2002), vol. 7(3):pp. 376–382, URL `http://dx.doi.org/10.1109/70.88147`

[Lig09] LIGGINS, Martin E.; HALL, David L. and LLINAS, James (Editors): *Handbook of Multisensor Data Fusion*, CRC Press, 2nd edn. (2009)

[Lim11] LIM, Gi Hyun; SUH, Il Hong and SUH, Hyowon: Ontology-Based Unified Robot Knowledge for Service Robots in Indoor Environments. *IEEE Transactions on Systems, Man, and Cybernetics – Part A: Systems and Humans* (2011), vol. 41(3):pp. 492–509

[Low04] LOWE, David G.: Distinctive Image Features from Scale-Invariant Keypoints. *International Journal of Computer Vision* (2004), vol. 60(2):pp. 91–110, URL `http://dx.doi.org/10.1023/B:VISI.0000029664.99615.94`

[Mac10a] MACHMER, T.; SWERDLOW, A.; KÜHN, B. and KROSCHEL, K.: Hierarchical, Knowledge-Oriented Opto-acoustic Scene Analysis for a Humanoid Robot and Man-Machine Interaction, in: *Proceedings of the IEEE International Conference on Robotics and Automation*, IEEE (2010), pp. 2389–2396

[Mac10b] MACHMER, Timo: *Generierung und Fusion von Umweltwissen für eine wissensbasierte Umwelterfassung*, Ph.D. thesis, Karlsruher Institut für Technologie (KIT), SierkeVerlag (2010)

[Mat05] MATSUI, Daisuke; MINATO, Takashi; MACDORMAN, Karl F. and ISHIGURO, Hiroshi: Generating Natural Motion in an Android by Mapping Human Motion, in: *IEEE/RSJ International Conference on Intelligent Robots and Systems (IROS 2005)*, IEEE (2005), pp. 3301–3308

[Mat08] MATHEWS, Zenon; BERMÚDEZ I BADIA, Sergi and VERSCHURE, Paul F.M.J.: Intelligent Motor Decision: From Selective Attention to a Bayesian World Model, in: *Intelligent Systems, 2008. IS '08. 4th International IEEE Conference*, vol. 3, IEEE (2008), pp. 4–8 – 4–13

[Mat10] MATHEWS, Zenon; BERMÚDEZ I BADIA, Sergi and VERSCHURE, Paul F.M.J.: *Intelligent Systems: From Theory to Practice*, chap. Action-Planning and Execution from Multimodal Cues: An Integrated Cognitive Model for Artificial Autonomous Systems, Springer Berlin Heidelberg (2010), pp. 479–497

[McC63] MCCARTHY, John: Situations, Actions, and Causal Laws, Tech. Rep., Stanford University (1963)

[MD09] MEYER-DELIUS, D.; PLAGEMANN, C. and BURGARD, W.: Probabilistic Situation Recognition for Vehicular Traffic Scenarios, in: *Proceedings of the 2009 IEEE international conference on Robotics and Automation*, IEEE (2009), pp. 4161–4166

[Men05] MENDOZA, Rogan and WILLIAMS, Mary-Anne: Ontology Based Object Categorisation for Robots, in: T. Meyer and M. Orgun (Editors) *Australasian Ontology Workshop, Conferences in Research and Practice in Information Technology (CRPIT)*, vol. 58, Australian Computer Society, Inc. (2005), pp. 61–67

[Moß10] MOSSGRABER, Jürgen; REINERT, Frank and VAGTS, Hauke: An architecture for a task-oriented surveillance system: a service and event based approach, in: *Proceedings of the Fifth International Conference on Systems*, IEEE (2010), pp. 146–151

[Muš94] MUŠICKI, Darko; EVANS, Robin J. and STANKOVIC, S.: Integrated probabilistic data association. *IEEE Transactions on Automatic Control* (1994), vol. 39(6):pp. 1237–1241

[Mus01] MUSSO, Christian; OUDJANE, Nadia and GLAND, Francois Le: Improving regularised particle filters, in: *Sequential Monte Carlo Methods in Practice*, Springer-Verlag (2001), pp. 247–271

[Muš04] MUŠICKI, Darko and EVANS, Robin J.: Joint integrated probabilistic data association: JIPDA. *IEEE Transactions on Aerospace and Electronic Systems* (2004), vol. 40(3):pp. 1093–1099

[Nis07] NISHIO, Shuichi; ISHIGURO, Hiroshi and HAGITA, Norihiro: *Humanoid Robots: New Developments*, chap. Geminoid: Teleoperated Android of an Existing Person, I-Tech Education and Publishing (2007), pp. 343–352

[Non10] NONAMI, K.; KENDOUL, F.; SUZUKI, S.; WANG, W. and NAKAZAWA, D.: *Autonomous Flying Robots*, Springer (2010)

[Num83] NUMERA, T. and STUBBERUD, A.R.: Gaussian sum approximation for non-linear fixed point prediction. *International Journal of Control* (1983), vol. 38:pp. 1047–1053

[ODE06a] Open Dynamics Engine (2006), URL `www.ode.org`

[Ode06b] ODELSON, Brian J.; LUTZ, Alexander and RAWLINGS, James B.: The Autocovariance Least-Squares Method for Estimating Covariances: Application to Model-Based Control of Chemical Reactors. *IEEE Transactions on Control Systems Technology* (2006), vol. 14(3):pp. 532–540

[Osw91] OSWALD, Jacques R.: *Diacritical Analysis of Systems: a Treatise on Information Theory*, Ellis Horwood Limited (1991)

[Pak13] PAK, Alexey; HUBER, Marco F. and BELKIN, Andrey: On weak distance between distributions in application to tracking, in: *Workshop on Sensor Data Fusion: Trends, Solutions, Applications (SDF)*, IEEE (2013)

[Pao94] PAO, Lucy Y.: Multisensor Multitarget Mixture Reduction Algorithms for Tracking. *AIAA Journal of Guidance, Control and Dynamics* (1994), vol. 17:pp. 1205–1211

[Pap08] PAPP, Z.; BROWN, C. and BARTELS, C.: World modeling for cooperative intelligent vehicles, in: *IEEE Intelligent Vehicles Symposium*, IEEE (2008), pp. 1050–1055

[Pla00] PLATANIOTIS, K.N. and HATZINAKOS, D.: *Advanced Signal Processing Handbook: Theory and Implementation: Theory and Implementation for Radar, Sonar, and Medical Imaging Real Time Systems*, chap. Gaussian mixtures and their applications to signal processing, CRC Press (2000), pp. 3–1–3–32

[Qui09] QUIGLEY, M.; CONLEY, K.; GERKEY, B.; FAUST, J.; FOOTE, T.; LEIBS, J.; WHEELER, R. and NG, A.: Ros: an open-source robot operating system, in: *IEEE International Conference on Robotics and Automation*, IEEE (2009)

[Rei79] REID, Donal B.: An Algorithm for Tracking Multiple Targets. *IEEE Transactions on Automatic Control* (1979), vol. AC-24(6):pp. 843–854

[Rem98] REMOLINA, E. and KUIPERS, B.: Towards a formalization of the spatial semantic hierarchy, in: *Proceedings of the 4th Symposium on Logical Formalizations Commonsense Reason* (1998)

[Rie97] RIEPP, Markus: *Wissensbasierte Parametrisierung von Operatorsequenzen*, Master's thesis, Karlsruhe University (TH) (1997)

[Riv95] RIVLIN, Ehud; DICKINSON, Sven and ROSENFELD, Azriel: Recognition by functional parts. *Computer Vision and Image Understanding* (1995), vol. September:pp. 267–274

[RL96] RONG LI, X. and BAR-SHALOM, Y.: Tracking in clutter with nearest neighbor filters: analysis and performance. *IEEE Transactions on Aerospace and Electronic Systems* (1996), vol. 32(3):pp. 995–1010

[Rob08] ROBOCUP: RoboCup web site (2008), URL `http://www.robocup.org`

[Rog60] ROGERS, David J. and TANIMOTO, Taffee T.: A Computer Program for Classifying Plants. *Science* (1960), vol. 132:pp. 1115–1118

[Rog03] ROGALLA, Oliver: *Abbildung von Benutzerdemonstrationen auf Variable Roboterkonfiguration*, Ph.D. thesis, Karlsruhe University (TH), GCA-Verlag (2003)

[Rog09] ROGERS, Albert J.; LIGHT, Edward D. and SMITH, Stephen W.: 3-D Ultrasound Guidance of Autonomous Robot for Location of Ferrous Shrapnel. *IEEE Transactions on Ultrasonics, Ferroelectrics and Frequency Control* (2009), vol. 56(7):pp. 1301–1303

[Roy04] ROY, Deb; HSIAO, Kai-Yuh and MAVRIDIS, Nikolaos: Mental Imagery for a Conversational Robot. *IEEE Transactions on Systems, Man, and Cybernetics, Part B: Cybernetics* (2004), vol. 34(3):pp. 1374–1383

[Rub12] RUBENSTEIN, Michael: Kilobot: A low cost scalable robot system for collective behaviors, in: *IEEE International Conference on Robotics and Automation (ICRA)*, IEEE (2012), pp. 3293–3298

[Run07] RUNNALLS, A. R.: Kullback-Leibler Approach to Gaussian Mixture Reduction. *IEEE Transactions on Aerospace and Electronic Systems* (2007), vol. 43:pp. 989–999

[Rus10] RUSSELL, Stuart and NORVIG, Peter (Editors): *Artificial intelligence*, Prentice Hall, Pearson Education, Inc. (2010)

[Sch11] SCHAUERTE, Boris; KÜHN, Benjamin; KROSCHEL, Kristian and STIEFELHAGEN, Rainer: Multimodal Saliency-Based Attention for Object-Based Scene Analysis, in: *Proceedings of the Conference on Intelligent Robots and Systems (IROS)*, IEEE, San Francisco, USA (2011)

[Seg11] SEGOR, F.; BÜRKLE, A.; KOLLMANN, M. and SCHÖNBEIN, R.: Instantaneous autonomous aerial reconnaissance for civil applications: A UAV based approach to support security and rescue forces, in: *Proceedings of the International Conference on Systems (ICONS)*, International Academy, Research, and Industry Association -IARIA-, St. Maarten, The Netherlands Antilles (2011), pp. 72–76

[Shi94] SHI, J. and TOMASI, C.: Good features to track, in: *Proceedings of the IEEE Computer Vision and Pattern Recognition*, IEEE (1994), pp. 593–600

[Sin02] SINGH, Push; LIN, Thomas; MUELLER, Erik T.; LIM, Grace; PERKINS, Travell and ZHU, Wan Li: Open Mind Common Sense: Knowledge Acquisition from the General Public, in: *On the Move to Meaningful Internet Systems 2002: CoopIS, DOA, and ODBASE*, vol. 2519 of *Lecture Notes in Computer Science*, Springer-Verlag (2002), pp. 1223–1237

[Sir07] SIRICHAROEN, Waralak V.: Ontologies and Object mode ls in Object Oriented Software Engineering. *IAENG International Journal of Computer Science* (2007), vol. 33(1):pp. 320–325

[Smi92] SMITH, A. F. M. and GELFAND, A. E.: Bayesian Statistics without Tears: A Sampling-Resampling Perspective. *The American Statistician* (1992), vol. 46(2):pp. 84–88

[Smi08] SMITS, R.; LAET, T. D.; CLAES, K.; SOETENS, P.; SCHUTTER, J. D. and BRUYNINCKX, H.: Orocos: A software framework for complex sensor-driven robot tasks. *IEEE Robotics and Automation Magazine* (2008)

[Sør57] SØRENSEN, Thorvald: A method of establishing groups of equal amplitude in plant sociology based on similarity of species and its application to analyses of the vegetation on Danish commons. *Biologiske Skrifter, Kongelige Danske Videnskabernes Selskab* (1957), vol. 5:pp. 1–34

[Sor71] SORENSON, H.W. and ALSPACH, D.L.: Recursive bayesian estimation using gaussian sums. *Automatica* (1971), vol. 7(4):pp. 465–479

[Spa02] SPAAN, M.T.J.; VLASSIS, N. and GROEN, F.C.A.: High level coordination of agents based on multiagent Markov decision processes with roles, in: *IROS'02 Workshop on Cooperative Robotics* (2002), pp. 66–73

[Ste98] STEINBERG, A.N.; BOWMAN, C.L. and WHITE, F.E.: Revisions to the JDL Model, in: *Joint NATO/IRIS Conference Proceedings,* The International Society for Optical Engineering (1998), reprinted in: Proceedings of the SPIE Sensor Fusion: Architectures, Algorithms, and Applications , Vol. 3719, 1999.

[Swe10] SWERDLOW, Alexej: *Audiovisuelle Signaturen für eine objektzentrierte Umwelterfassung,* Ph.D. thesis, Karlsruhe Institut of Technology (KIT), SierkeVerlag (2010)

[Tel08] TELLEZ, Ricardo; FERRO, Francesco; GARCIA, Sergio; GOMEZ, Esteban; JORGE, Enric; MORA, Dario; PINYOL, Daniel; POYATOS, Joan; TORRES, Oriol; VELAZQUEZ, Jorge and FACONTI, Davide: Reem-B: an autonomous lightweight human-size humanoid robot, in: *8th IEEE-RAS International Conference on Humanoid Robots,* IEEE (2008), pp. 462–468

[Ten09] TENORTH, M. and BEETZ, M.: KNOWROB – knowledge processing for autonomous personal robots, in: *Proceedings of the IEEE/RSJ International Conference on Intelligent Robots and Systems,* IEEE (2009), pp. 4261–4266

[Thr03] THRUN, Sebastian: Learning Occupancy Grid Maps with Forward Sensor Models. *Autonomous Robots* (2003), vol. 15(2):pp. 111–127

[Thr05] THRUN, Sebastian; BURGARD, Wolfram and FOX, Dieter: *Probabilistic Robotics,* Intelligent Robotics and Autonomous Agents, The MIT Press (2005)

[TOS10] TOSY ROBOTICS: TOSY Ping Pong Playing Robot (TOPIO) (2010), URL `http://topio.tosy.com/`

[Vec05] VECNA TECHNOLOGIES, INC.: The BEAR™ (2005)

[Ver05] VERMAAK, J.; MASKELL, S. and BRIERS, M.: A unifying framework for multi-target tracking and existence, in: *8th International Conference on Information Fusion,* vol. 1, IEEE (2005)

[Vio01] VIOLA, Paul and JONES, Michael: Robust Real-time Object Detection, in: *International Journal of Computer Vision*, Kluwer Academic Publishers (2001), pp. 137–154

[Vla01] VLASSIS, N.; KOK, J.R.; SPAAN, M.T.J.; TERWIJN, B.; BOASSON, M.; GROEN, F.C.A.; SPOELDER, H.J.W. and BAL, H.E.: A framework for maintaining a shared world model in dynamic environments between differentiated embedded systems and allowing interaction with human supervisors, in: *Proceedings of the 2nd PROGRESS Workshop on Embedded Systems*, Veldhoven, The Netherlands (2001)

[Vo06] VO, Ba-Ngu and MA, Wing-Kin: The gaussian mixture probability hypothesis density filter. *IEEE TRANSACTIONS ON SIGNAL PROCESSING* (2006), vol. 54:pp. 4091–4104

[Wes93] WEST, Mike: Approximating Posterior Distributions by Mixture. *Journal of the Royal Statistical Society. Series B (Methodological)* (1993), vol. 55:pp. 409–422

[Whi91] WHITE, Franklin E.: *DATA FUSION LEXICON*, The DATA FUSION SUBPANEL of the Joint Directors of Laboratories, Technical Panel for C3 (1991)

[Wil03a] WILLIAMS, Jason L.: *Gaussian Mixture Reduction for Tracking Multiple Maneuvering Targets in Clutter*, Master's thesis, Graduate School of Engineering and Management, Air Force Institute of Technology, Air University, USA (2003)

[Wil03b] WILLIAMS, Jason L. and MAYBECK, Peter S.: Cost-Function-Based Gaussian Mixture Reduction for Target Tracking, in: *Proceedings of the 6th International Conference on Information Fusion (Fusion)*, vol. 2, IEEE (2003), pp. 1047–1054

[Win13] WINKLER, Jan; TENORTH, Moritz; BOZCUOGLU, Asil Kaan and BEETZ, Michael: CRAMm – Memories for Robots Performing Everyday Manipulation Activities. *Second Annual Conference on Advances in Cognitive Systems* (2013), vol. 3:pp. 47–66

[Wol81] WOLDA, Henk: Similarity indices, sample size and diversity. *Oecologia* (1981), vol. 50(3):pp. 296–302

[Yan10] YANG, Zhengwei: A study of Land Cover Change Detection with Tanimoto Distance, in: *Association of American Geographers Annual Meeting*, USDA/NASS, Washington, DC, USA (2010)

[Zam11] ZAMANIFAR, Kamran and ALAMIYAN, Farinaz: A New Similarity Measure for Instance Data Matching, in: *Proceedings of the International Conference on Computer Communication and Management* (2011)

[Zee97] ZEEVI, A.S. and MEIR, R.: Density estimation through convex combination of densities: Approximation and estimation bounds. *Neural Networks* (1997), vol. 10(1):pp. 99–109

# List of Figures

# List of Tables

# Acknowledgment

The presented dissertation was written as a result of my work as a research fellow at Vision and Fusion Laboratory (IES), Institute for Anthropomatics, University of Karlsruhe, which was transformed meanwhile into Karlsruhe Institute of Technology (KIT). I would like to thank the laboratory chair Prof. Dr.-Ing. Jürgen Beyerer for brilliant scientific and administrative supervision, as well as for the chance to participate in this research.

This work was supported by the German Science Foundation (Deutsche Forschungsgemeinschaft, DFG) within the Collaborative Research Center (Sonderforschungsbereich, SFB) 588 "Humanoid Robots – Learning and Cooperating Multimodal Robots" and Fraunhofer Institute of Optronics, System Technologies and Image Exploitation (IOSB) institute.

I am thankful to Prof. Dr.-Ing. Tamim Asfour from High Performance Humanoid Technologies Lab, Institute for Anthropomatics, KIT for giving numerous informative talks and carrying duty of second reviewer. Also, I feel gratitude to all professors and doctors who have helped me throughout the work, especially to Dr. Alexey Pak (for invaluable scientific help, insights and thorough reading of the manuscript) and Prof. Dr.-Ing. Kristian Kroschel. I have to mention IES secretary Gaby Gross and SFB 588 secretary Isabelle Wappler for constant and comprehensive assistance with all those formalities.

I would like to thank all my colleagues, friends and relatives for the help and support throughout my work. I can mention several names that I recall at the moment (most of them are already bearing "Dr." in front of the name or will do so soon): Ioana Gheţa, Michael Grinberg, Alexander Kasper, Michael Heizmann, Stefan Werling, Miriam Schleipen, Jennifer Sander, Michael Mai, Robin Gruna, Christian Frese, Marco Huber, Alexej Swerdlow, Michael Teutsch, Miro Taphanel, Masoud Roschani, Achim Kuwertz, Chettapong Janya-Anurak, and Sebastian Höfer. I give my special thanks to my mother Ivetta Kim for the infinite support and to my friend Dr. Mikhail Tentyukov for the final reading of the manuscript.

Karlsruhe, October 4, 2016 Andrey Belkin

## Karlsruher Schriftenreihe zur Anthropomatik (ISSN 1863-6489)

Herausgeber: Prof. Dr.-Ing. Jürgen Beyerer

**Band 1** Jürgen Geisler
**Leistung des Menschen am Bildschirmarbeitsplatz.** 2006
ISBN 3-86644-070-7

**Band 2** Elisabeth Peinsipp-Byma
**Leistungserhöhung durch Assistenz in interaktiven Systemen zur Szenenanalyse.** 2007
ISBN 978-3-86644-149-1

**Band 3** Jürgen Geisler, Jürgen Beyerer (Hrsg.)
**Mensch-Maschine-Systeme.** 2010
ISBN 978-3-86644-457-7

**Band 4** Jürgen Beyerer, Marco Huber (Hrsg.)
**Proceedings of the 2009 Joint Workshop of Fraunhofer IOSB and Institute for Anthropomatics, Vision and Fusion Laboratory.** 2010
ISBN 978-3-86644-469-0

**Band 5** Thomas Usländer
**Service-oriented design of environmental information systems.** 2010
ISBN 978-3-86644-499-7

**Band 6** Giulio Milighetti
**Multisensorielle diskret-kontinuierliche Überwachung und Regelung humanoider Roboter.** 2010
ISBN 978-3-86644-568-0

**Band 7** Jürgen Beyerer, Marco Huber (Hrsg.)
**Proceedings of the 2010 Joint Workshop of Fraunhofer IOSB and Institute for Anthropomatics, Vision and Fusion Laboratory.** 2011
ISBN 978-3-86644-609-0

**Band 8** Eduardo Monari
**Dynamische Sensorselektion zur auftragsorientierten Objektverfolgung in Kameranetzwerken.** 2011
ISBN 978-3-86644-729-5

Die Bände sind unter www.ksp.kit.edu als PDF frei verfügbar oder als Druckausgabe bestellbar.

**Band 9** Thomas Bader
**Multimodale Interaktion in Multi-Display-Umgebungen.** 2011
ISBN 3-86644-760-8

**Band 10** Christian Frese
**Planung kooperativer Fahrmanöver für kognitive Automobile.** 2012
ISBN 978-3-86644-798-1

**Band 11** Jürgen Beyerer, Alexey Pak (Hrsg.)
**Proceedings of the 2011 Joint Workshop of Fraunhofer IOSB and Institute for Anthropomatics, Vision and Fusion Laboratory.** 2012
ISBN 978-3-86644-855-1

**Band 12** Miriam Schleipen
**Adaptivität und Interoperabilität von Manufacturing Execution Systemen (MES).** 2013
ISBN 978-3-86644-955-8

**Band 13** Jürgen Beyerer, Alexey Pak (Hrsg.)
**Proceedings of the 2012 Joint Workshop of Fraunhofer IOSB and Institute for Anthropomatics, Vision and Fusion Laboratory.** 2013
ISBN 978-3-86644-988-6

**Band 14** Hauke-Hendrik Vagts
**Privatheit und Datenschutz in der intelligenten Überwachung: Ein datenschutzgewährendes System, entworfen nach dem „Privacy by Design" Prinzip.** 2013
ISBN 978-3-7315-0041-4

**Band 15** Christian Kühnert
**Data-driven Methods for Fault Localization in Process Technology.** 2013
ISBN 978-3-7315-0098-8

**Band 16** Alexander Bauer
**Probabilistische Szenenmodelle für die Luftbildauswertung.** 2014
ISBN 978-3-7315-0167-1

**Band 17** Jürgen Beyerer, Alexey Pak (Hrsg.)
**Proceedings of the 2013 Joint Workshop of Fraunhofer IOSB and Institute for Anthropomatics, Vision and Fusion Laboratory.** 2014
ISBN 978-3-7315-0212-8

---

**Band 18** Michael Teutsch
**Moving Object Detection and Segmentation for Remote Aerial Video Surveillance.** 2015
ISBN 978-3-7315-0320-0

**Band 19** Marco Huber
**Nonlinear Gaussian Filtering: Theory, Algorithms, and Applications.** 2015
ISBN 978-3-7315-0338-5

**Band 20** Jürgen Beyerer, Alexey Pak (Hrsg.)
**Proceedings of the 2014 Joint Workshop of Fraunhofer IOSB and Institute for Anthropomatics, Vision and Fusion Laboratory.** 2014
ISBN 978-3-7315-0401-6

**Band 21** Todor Dimitrov
**Permanente Optimierung dynamischer Probleme der Fertigungssteuerung unter Einbeziehung von Benutzerinteraktionen.** 2015
ISBN 978-3-7315-0426-9

**Band 22** Benjamin Kühn
**Interessengetriebene audiovisuelle Szenenexploration.** 2016
ISBN 978-3-7315-0457-3

**Band 23** Yvonne Fischer
**Wissensbasierte probabilistische Modellierung für die Situationsanalyse am Beispiel der maritimen Überwachung.** 2016
ISBN 978-3-7315-0460-3

**Band 24** Jürgen Beyerer, Alexey Pak (Hrsg.)
**Proceedings of the 2015 Joint Workshop of Fraunhofer IOSB and Institute for Anthropomatics, Vision and Fusion Laboratory.** 2016
ISBN 978-3-7315-0519-8

**Band 25** Pascal Birnstill
**Privacy-Respecting Smart Video Surveillance Based on Usage Control Enforcement.** 2016
ISBN 978-3-7315-0538-9

**Band 26** Philipp Woock
**Umgebungskartenschätzung aus Sidescan-Sonardaten für ein autonomes Unterwasserfahrzeug.** 2016
ISBN 978-3-7315-0541-9

**Band 27** Janko Petereit
**Adaptive State × Time Lattices: A Contribution to Mobile Robot Motion Planning in Unstructured Dynamic Environments.** 2017
ISBN 978-3-7315-0580-8

**Band 28** Erik Ludwig Krempel
**Steigerung der Akzeptanz von intelligenter Video-überwachung in öffentlichen Räumen.** 2017
ISBN 978-3-7315-0598-3

**Band 29** Jürgen Moßgraber
**Ein Rahmenwerk für die Architektur von Frühwarnsystemen.** 2017
ISBN 978-3-7315-0638-6

**Band 30** Andrey Belkin
**World Modeling for Intelligent Autonomous Systems.** 2017
ISBN 978-3-7315-0641-6